THE PERSE PREPARATORY SCHOOL

Perse Preparatory School
R03191M0345

D1354890

EYEWITNESS VISUAL DICTIONARIES

THE VISUAL DICTIONARY *of*
CHEMISTRY

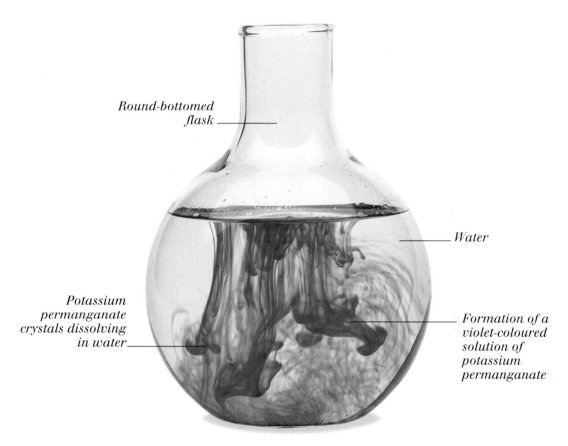

Round-bottomed
flask

Water

Potassium
permanganate
crystals dissolving
in water

Formation of a
violet-coloured
solution of
potassium
permanganate

DISSOLVING POTASSIUM PERMANGANATE IN WATER

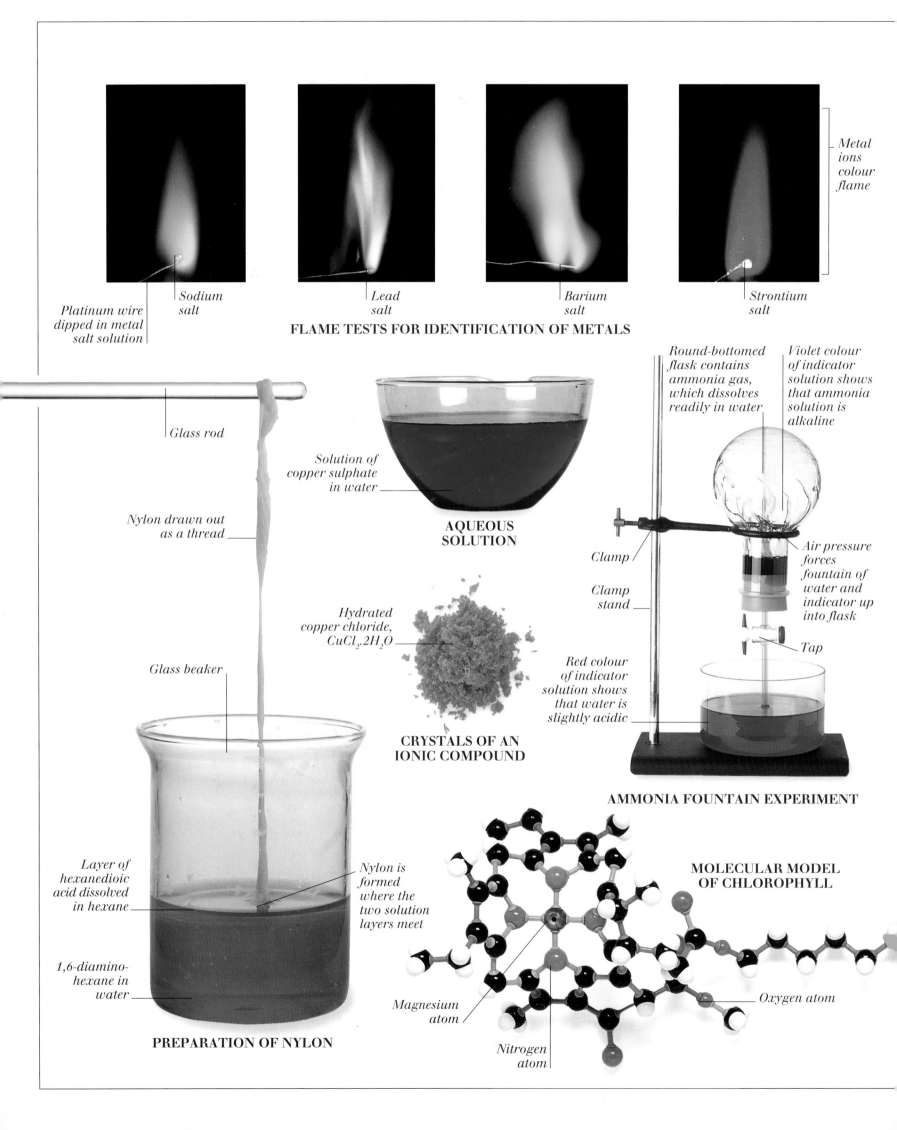

Metal ions colour flame

Sodium salt

Lead salt

Barium salt

Strontium salt

Platinum wire dipped in metal salt solution

FLAME TESTS FOR IDENTIFICATION OF METALS

Glass rod

Nylon drawn out as a thread

Glass beaker

Layer of hexanedioic acid dissolved in hexane

Nylon is formed where the two solution layers meet

1,6-diamino-hexane in water

PREPARATION OF NYLON

Solution of copper sulphate in water

AQUEOUS SOLUTION

Hydrated copper chloride, $CuCl_2.2H_2O$

CRYSTALS OF AN IONIC COMPOUND

Round-bottomed flask contains ammonia gas, which dissolves readily in water

Violet colour of indicator solution shows that ammonia solution is alkaline

Clamp

Clamp stand

Air pressure forces fountain of water and indicator up into flask

Tap

Red colour of indicator solution shows that water is slightly acidic

AMMONIA FOUNTAIN EXPERIMENT

MOLECULAR MODEL OF CHLOROPHYLL

Magnesium atom

Oxygen atom

Nitrogen atom

EYEWITNESS VISUAL DICTIONARIES

THE VISUAL DICTIONARY *of*
CHEMISTRY

written by
Jack Challoner

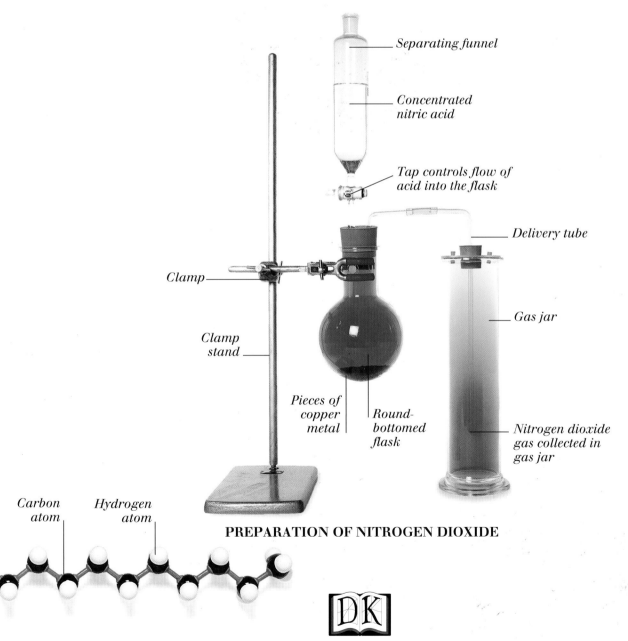

Separating funnel

Concentrated nitric acid

Tap controls flow of acid into the flask

Delivery tube

Clamp

Gas jar

Clamp stand

Pieces of copper metal

Round-bottomed flask

Nitrogen dioxide gas collected in gas jar

Carbon atom

Hydrogen atom

PREPARATION OF NITROGEN DIOXIDE

DK

DORLING KINDERSLEY

LONDON • NEW YORK • SYDNEY • MOSCOW

A DORLING KINDERSLEY BOOK

ART EDITOR Simon Murrell
PROJECT EDITOR Mukul Patel
EDITOR Des Reid
DESIGN ASSISTANT Claire Naylor

DEPUTY ART DIRECTOR Tina Vaughan
MANAGING EDITOR Sean Moore
SENIOR ART EDITOR Tracy Hambleton-Miles
SENIOR EDITOR Louise Candlish

PHOTOGRAPHY Dean Belcher, Andy Crawford, Tim Ridley
ILLUSTRATIONS Chris Lyon

PRODUCTION Stephen Stuart

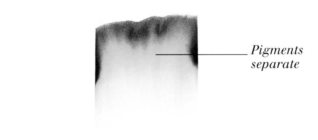

Pigments separate

*Filter paper*_____   *Watch glass*

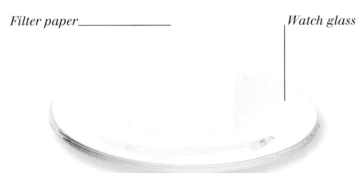

PAPER CHROMATOGRAPHY

FIRST PUBLISHED IN GREAT BRITAIN IN 1996
BY DORLING KINDERSLEY LIMITED,
9 HENRIETTA STREET, LONDON WC2E 8PS

COPYRIGHT © 1996 DORLING KINDERSLEY LIMITED, LONDON
TEXT COPYRIGHT © 1996 JACK CHALLONER

Reprinted 1997

ALL RIGHTS RESERVED. NO PART OF THIS PUBLICATION MAY BE REPRODUCED,
STORED IN A RETRIEVAL SYSTEM, OR TRANSMITTED IN ANY FORM OR BY ANY MEANS, ELECTRONIC,
MECHANICAL, PHOTOCOPYING, RECORDING OR OTHERWISE, WITHOUT THE PRIOR
WRITTEN PERMISSION OF THE COPYRIGHT OWNER.

A CIP CATALOGUE RECORD FOR THIS BOOK IS AVAILABLE FROM THE BRITISH LIBRARY

ISBN 0 7513 1061 1

REPRODUCED BY COLOURSCAN, SINGAPORE
PRINTED AND BOUND BY ARNOLDO MONDADORI, VERONA, ITALY

WARNING
Do not attempt to carry out any of the experiments shown in this book without the assistance of a trained chemist or chemistry teacher. Some of the experiments produce large amounts of heat or poisonous gases.

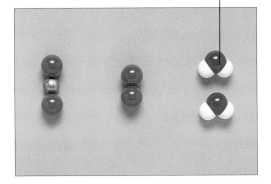

Molecule of reaction product

MOLECULAR MODEL OF REACTION

Brightly coloured crystals of transition metal compound

Contents

IONIC COMPOUND

Hydrated crystals

COBALT(II) CHLORIDE

Silver crystals

DISPLACEMENT REACTION

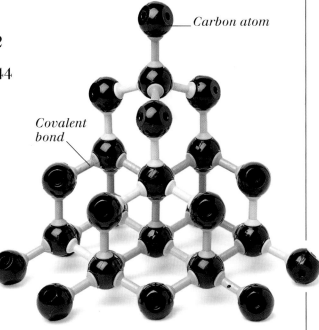

Conical flask

Potassium chromate solution

REVERSIBLE REACTION

Carbon atom

Covalent bond

MOLECULAR STRUCTURE OF DIAMOND

Elements and compounds

CHEMISTRY IS THE STUDY OF MATTER. All ordinary matter consists of tiny units called **atoms** (see pp. 10-11). An **element** is a substance that contains atoms of one type only. However, pure elements are rarely found in nature – they are nearly always combined with other elements. A **compound** is a substance in which the atoms of two or more elements are combined in definite proportions. The atoms in a compound are often bound together in units called **molecules**. For example, each molecule of the compound ammonia, NH_3, consists of one atom of nitrogen, N, bound to three of hydrogen, H. Atoms interact with each other during **chemical reactions**, making or breaking bonds to form new substances. The **products** of a reaction often have very different properties to the original **reactants**. For example, iron, a magnetic element, reacts with the yellow element sulphur to produce iron(II) sulphide, which is neither magnetic nor yellow. Similarly, the compound mercury(II) oxide is an orange powder – very different from its constituent elements.

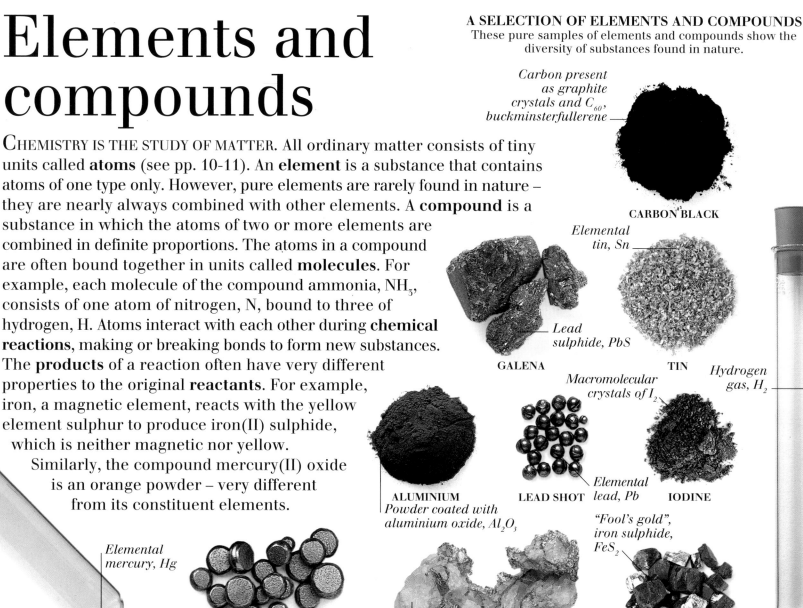

A SELECTION OF ELEMENTS AND COMPOUNDS
These pure samples of elements and compounds show the diversity of substances found in nature.

Carbon present as graphite crystals and C_{60}, buckminsterfullerene

CARBON BLACK

Elemental tin, Sn

Lead sulphide, PbS

GALENA

TIN

Hydrogen gas, H_2

Macromolecular crystals of I_2

ALUMINIUM
Powder coated with aluminium oxide, Al_2O_3

LEAD SHOT
Elemental lead, Pb

IODINE

"Fool's gold", iron sulphide, FeS_2

Elemental mercury, Hg

MERCURY

Elemental nickel, Ni
NICKEL

Quartz, silicon dioxide, SiO_2

Veins of elemental gold, Au

GOLD AND QUARTZ CRYSTAL

IRON PYRITES

HYDROGEN

MOLECULES

MOLECULAR MODELS
Many compounds exist as individual molecules. Models of molecules can help us to understand and predict chemical reactions. Space-filling models show how the atoms that make up a molecule overlap. Ball and stick models show the bonds and bond angles between the atoms.

Oxygen atom

Hydrogen atom

Carbon atom

Space-filling model

ETHANOL, C_2H_5OH

Oxygen atom

Bond between atoms

Hydrogen atom

Carbon atom

Ball and stick model

Oxygen atom

Hydrogen atom

Space-filling model

Oxygen atom

Bond between atoms

Hydrogen atom

Bond angle 105°

Ball and stick model

WATER, H_2O

Nitrogen atom

Hydrogen atom

Space-filling model

Bond between atoms

Hydrogen atom

Nitrogen atom

Bond angle 107°

Ball and stick model

AMMONIA, NH_3

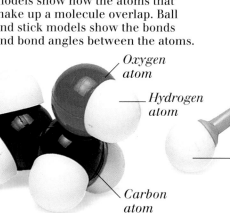

PREPARATION OF IRON(II) SULPHIDE

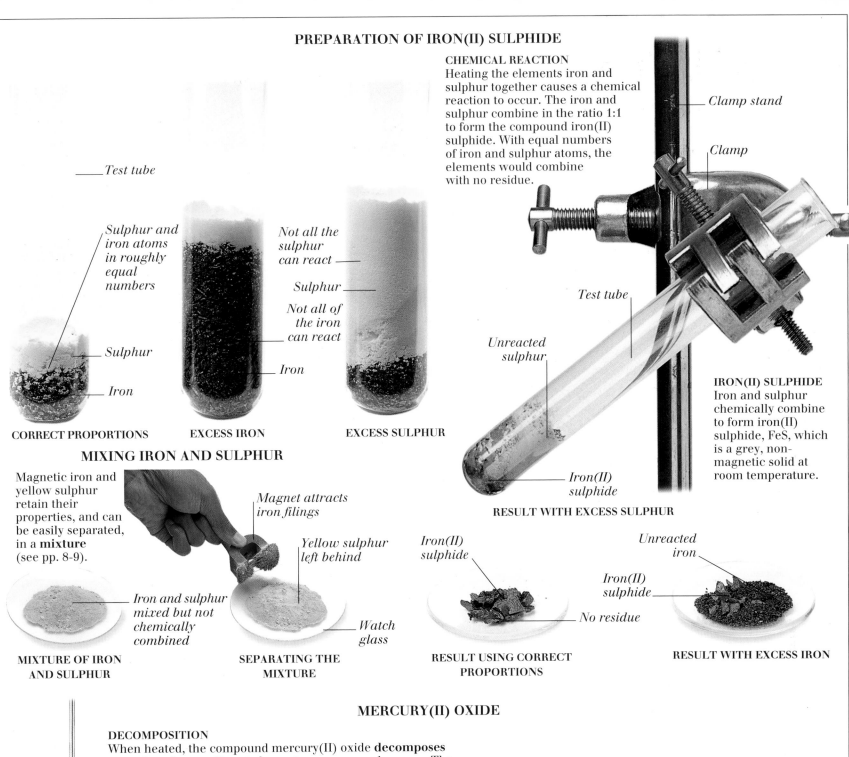

CHEMICAL REACTION
Heating the elements iron and sulphur together causes a chemical reaction to occur. The iron and sulphur combine in the ratio 1:1 to form the compound iron(II) sulphide. With equal numbers of iron and sulphur atoms, the elements would combine with no residue.

Test tube

Sulphur and iron atoms in roughly equal numbers

Sulphur

Iron

CORRECT PROPORTIONS

Not all the sulphur can react

Sulphur

Iron

EXCESS IRON

Not all of the iron can react

Sulphur

EXCESS SULPHUR

Clamp stand

Clamp

Test tube

Unreacted sulphur

Iron(II) sulphide

RESULT WITH EXCESS SULPHUR

IRON(II) SULPHIDE
Iron and sulphur chemically combine to form iron(II) sulphide, FeS, which is a grey, non-magnetic solid at room temperature.

MIXING IRON AND SULPHUR

Magnetic iron and yellow sulphur retain their properties, and can be easily separated, in a **mixture** (see pp. 8-9).

Magnet attracts iron filings

Yellow sulphur left behind

Iron and sulphur mixed but not chemically combined

Watch glass

MIXTURE OF IRON AND SULPHUR

SEPARATING THE MIXTURE

Iron(II) sulphide

No residue

RESULT USING CORRECT PROPORTIONS

Unreacted iron

Iron(II) sulphide

RESULT WITH EXCESS IRON

MERCURY(II) OXIDE

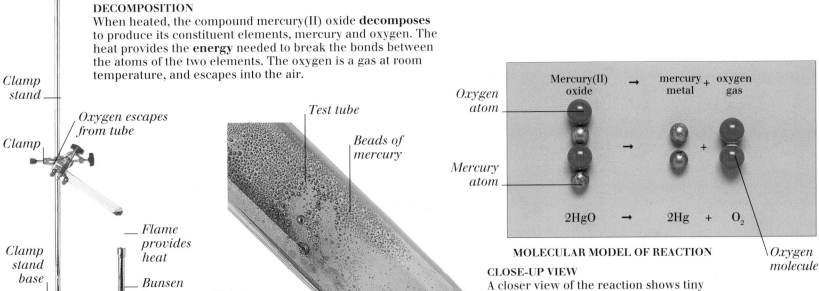

DECOMPOSITION
When heated, the compound mercury(II) oxide **decomposes** to produce its constituent elements, mercury and oxygen. The heat provides the **energy** needed to break the bonds between the atoms of the two elements. The oxygen is a gas at room temperature, and escapes into the air.

Clamp stand

Oxygen escapes from tube

Clamp

Clamp stand base

Flame provides heat

Bunsen burner

Test tube

Beads of mercury

This form of mercury(II) oxide is an orange powder

Mercury(II) oxide → mercury metal + oxygen gas

Oxygen atom

Mercury atom

$2HgO$ → $2Hg$ + O_2

MOLECULAR MODEL OF REACTION

Oxygen molecule

CLOSE-UP VIEW
A closer view of the reaction shows tiny beads of the mercury metal produced. The models above present a molecular view of the reaction, while the equation summarizes the reaction symbolically.

Mixtures

A MIXTURE CONTAINS TWO or more pure substances (**elements** or **compounds**), which may be solids, liquids, or gases. For example, air is a **mixture** of gases, cement is a mixture of solids, and sea water is a mixture of solids, liquids, and gases. A **solution** is a common type of mixture, consisting of a **solute** (often a solid) mixed evenly with a **solvent** (usually liquid). When the solvent is water, the solute particles are usually **ions**. Other types of mixture include **colloids**, like milk, in which the dispersed particles are slightly larger than ions, and **suspensions**, in which they are larger still. Because the substances making up a mixture are not chemically combined (see pp. 16-17), they can be separated easily. **Chromatography** is used to separate mixtures for analysis, for example in breathalysers. A technique called **filtration** is used to separate suspensions such as muddy water. Solutions may be separated by **distillation**, in which the solvent is boiled off and collected, and the solute is left behind. If both the solute and the solvent are liquids, then a technique called fractional distillation is used (see pp. 50-51).

AIR AS A MIXTURE

The coloured balls in this column represent the proportions of gases in dry air. Usually, air also contains water vapour and dust particles.

Nitrogen (white) makes up 78% of the air

Oxygen (orange) makes up 21% of the air

Argon (red) makes up 0.93% of the air

Carbon dioxide (black) makes up 0.03% of the air

SOLUTIONS

Nickel(II) nitrate is a solid at room temperature. It dissolves well in water to give a green coloured **aqueous solution**.

100 ml beaker

Water (solvent)

Solid dissolves in water

Nickel(II) nitrate (solute)

Aqueous solution of nickel(II) nitrate forms

SOLUTION OF NICKEL(II) NITRATE IN WATER

Particle in solution

Water molecule

Particles break away from solid

MICROSCOPIC VIEW
When a solid dissolves in a liquid solvent such as water, the particles of the solid break away and mix evenly and thoroughly with particles of the liquid.

PAPER CHROMATOGRAPHY

Ink from a felt-tip pen is dissolved in alcohol in a glass dish. The alcohol soaks into the absorbent filter paper, carrying the ink with it. Coloured ink is a mixture of several pigments, which bind to the paper to different extents. Those pigments that bind loosely move more quickly up the paper than the others, and so the ink separates into its constituent pigments.

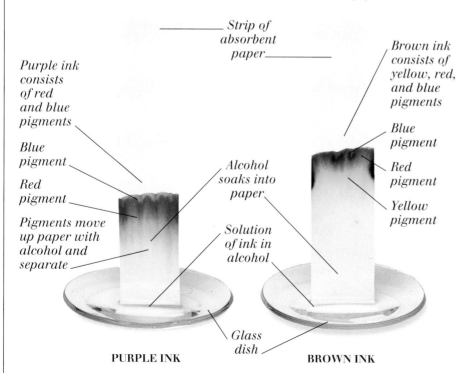

Strip of absorbent paper

Purple ink consists of red and blue pigments

Blue pigment

Red pigment

Pigments move up paper with alcohol and separate

Alcohol soaks into paper

Solution of ink in alcohol

Brown ink consists of yellow, red, and blue pigments

Blue pigment

Red pigment

Yellow pigment

Glass dish

PURPLE INK

BROWN INK

GAS CHROMATOGRAPHY

The sample for analysis is vaporized and carried through a granulated solid by a moving stream of an inert gas such as helium. Different parts of the sample travel at different rates through the solid, and can be identified by a sensitive detector.

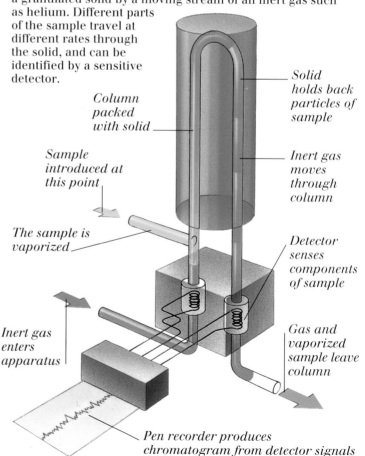

Column packed with solid

Sample introduced at this point

The sample is vaporized

Inert gas enters apparatus

Solid holds back particles of sample

Inert gas moves through column

Detector senses components of sample

Gas and vaporized sample leave column

Pen recorder produces chromatogram from detector signals

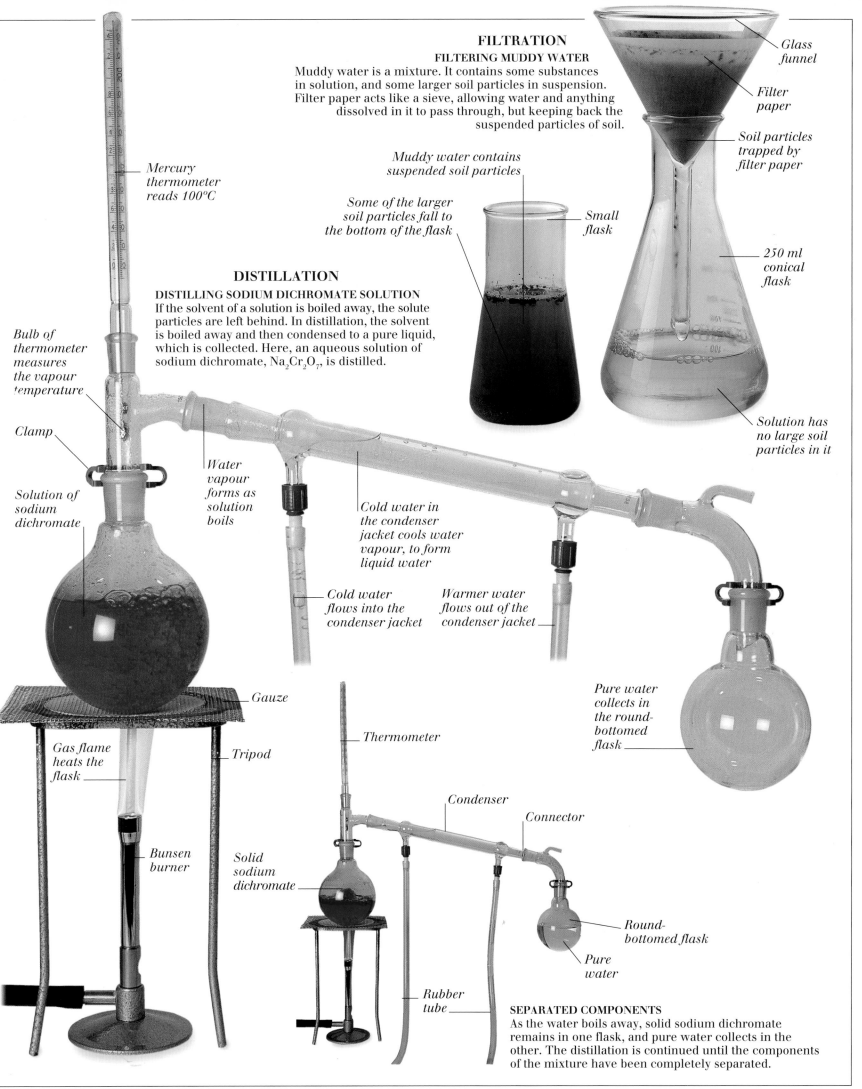

FILTRATION

FILTERING MUDDY WATER

Muddy water is a mixture. It contains some substances in solution, and some larger soil particles in suspension. Filter paper acts like a sieve, allowing water and anything dissolved in it to pass through, but keeping back the suspended particles of soil.

Glass funnel

Filter paper

Soil particles trapped by filter paper

Muddy water contains suspended soil particles

Some of the larger soil particles fall to the bottom of the flask

Small flask

250 ml conical flask

Solution has no large soil particles in it

DISTILLATION

DISTILLING SODIUM DICHROMATE SOLUTION

If the solvent of a solution is boiled away, the solute particles are left behind. In distillation, the solvent is boiled away and then condensed to a pure liquid, which is collected. Here, an aqueous solution of sodium dichromate, $Na_2Cr_2O_7$, is distilled.

Mercury thermometer reads 100°C

Bulb of thermometer measures the vapour temperature

Clamp

Solution of sodium dichromate

Water vapour forms as solution boils

Cold water in the condenser jacket cools water vapour, to form liquid water

Cold water flows into the condenser jacket

Warmer water flows out of the condenser jacket

Gauze

Gas flame heats the flask

Tripod

Thermometer

Pure water collects in the round-bottomed flask

Bunsen burner

Solid sodium dichromate

Condenser

Connector

Round-bottomed flask

Pure water

Rubber tube

SEPARATED COMPONENTS

As the water boils away, solid sodium dichromate remains in one flask, and pure water collects in the other. The distillation is continued until the components of the mixture have been completely separated.

Atoms and molecules

EVERY ATOM CONTAINS AN equal number of **electrically charged protons** and **electrons**, and a number of uncharged **neutrons**. Neutrons and the positively charged protons are found in the central **nucleus**. The nucleus is surrounded by negatively charged electrons, which take part in chemical bonding (see pp. 16-17). Each **element** has a unique **atomic number** – the number of protons in its **atoms** – though the number of neutrons varies between different **isotopes** of the element. An atom's mass may be given simply as the total number of neutrons and protons, since these particles have nearly equal masses, far greater than that of an electron. The **relative atomic mass (RAM)** is a more precise measure, based on the accurately determined atomic mass of a carbon isotope. The sum of the RAMs of the elements making up a **compound** is called the **relative molecular mass (RMM)**. One **mole** of a substance has the same mass in grams as its RAM or RMM. The mole is a useful unit, because it specifies a fixed number of atoms, **ions**, or **molecules**.

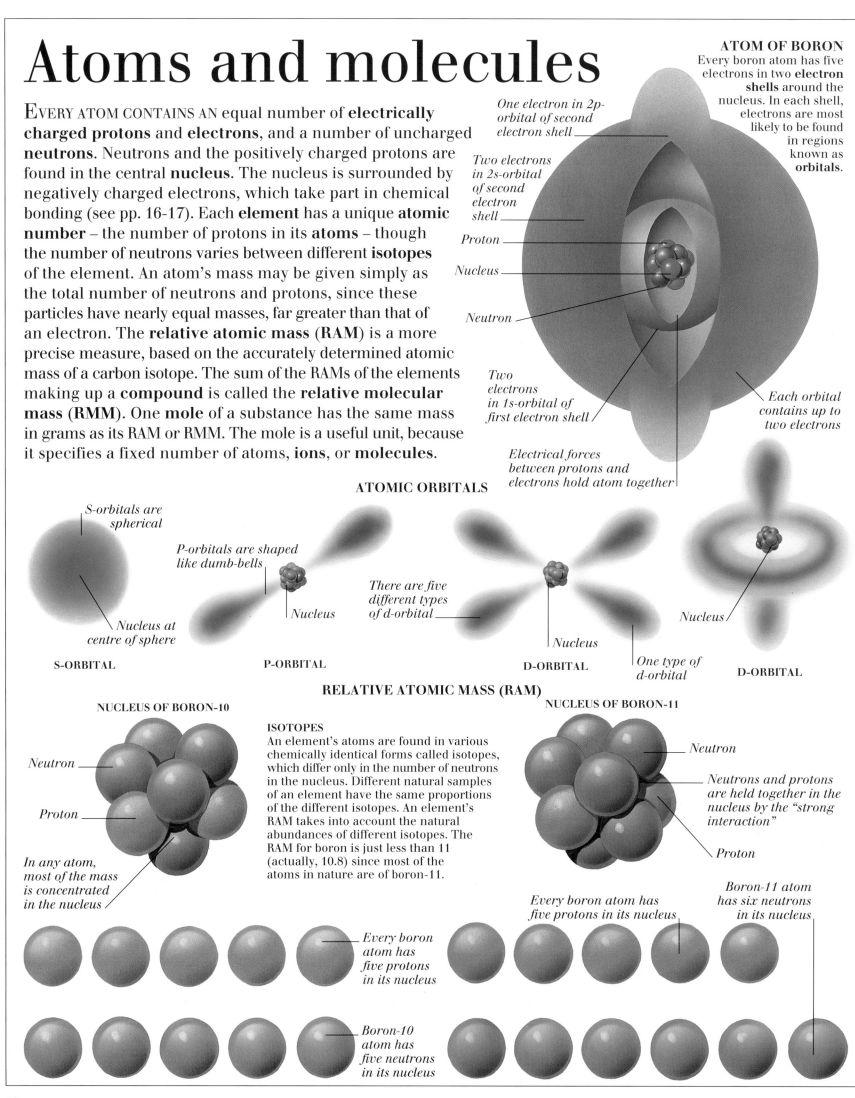

ATOM OF BORON
Every boron atom has five electrons in two **electron shells** around the nucleus. In each shell, electrons are most likely to be found in regions known as **orbitals**.

One electron in 2p-orbital of second electron shell

Two electrons in 2s-orbital of second electron shell

Proton

Nucleus

Neutron

Two electrons in 1s-orbital of first electron shell

Electrical forces between protons and electrons hold atom together

Each orbital contains up to two electrons

ATOMIC ORBITALS

S-orbitals are spherical

Nucleus at centre of sphere

S-ORBITAL

P-orbitals are shaped like dumb-bells

Nucleus

P-ORBITAL

There are five different types of d-orbital

Nucleus

D-ORBITAL

One type of d-orbital

Nucleus

D-ORBITAL

RELATIVE ATOMIC MASS (RAM)

NUCLEUS OF BORON-10

Neutron

Proton

In any atom, most of the mass is concentrated in the nucleus

NUCLEUS OF BORON-11

Neutron

Neutrons and protons are held together in the nucleus by the "strong interaction"

Proton

ISOTOPES
An element's atoms are found in various chemically identical forms called isotopes, which differ only in the number of neutrons in the nucleus. Different natural samples of an element have the same proportions of the different isotopes. An element's RAM takes into account the natural abundances of different isotopes. The RAM for boron is just less than 11 (actually, 10.8) since most of the atoms in nature are of boron-11.

Every boron atom has five protons in its nucleus

Boron-11 atom has six neutrons in its nucleus

Every boron atom has five protons in its nucleus

Boron-10 atom has five neutrons in its nucleus

GAS MOLAR VOLUME

One mole of any gas at STP would fill up more than 22 of these bottles

ONE LITRE BOTTLE

BOX CONTAINING ONE MOLE OF GAS

GAS VOLUME AT STP
One mole of any gas at standard temperature and pressure (**STP**) always occupies 22.4 litres of space. Although the number of particles (atoms or molecules) making up one mole of a gas is extremely large, each particle is very tiny. This means that the volume of a gas only depends upon the number of particles present, and not on the size of each particle. The box and the bottle (left) give an idea of the molar volume of any gas at STP.

MOLAR MASSES

ONE MOLE OF COPPER
Copper has an RAM of 64.4, so the molar mass of copper is 64.4 grams. The number of atoms present is 6.02 x 10²³.

126.9 grams of iodine (one mole)

64.4 grams of copper (one mole)

Copper is a metallic element

ONE MOLE OF IODINE
The element iodine has an RAM of 126.9. The molar mass of iodine is 126.9 grams. The number of atoms, ions, or molecules in one mole of any substance is 6.02 x 10²³ – a number known as Avogadro's constant.

Iodine is a violet solid at room temperature

PREPARING A 0.1 M SOLUTION OF COBALT CHLORIDE

Plastic stopper

0.1 MOLAR SOLUTION OF COBALT CHLORIDE
Enough water is mixed thoroughly with 0.1 mole of cobalt chloride (below left) to make exactly one litre of **solution**. The cobalt chloride dissolves to form a 0.1 molar (0.1M) solution. This is the **concentration** of the solution, sometimes known as its molarity.

Volumetric flask

Neck of flask is narrow, so that it may be accurately filled

Etched mark on flask indicates one litre capacity

Solution of cobalt chloride

0.1 MOLE OF COBALT CHLORIDE
The RMM of hydrated cobalt chloride, $CoCl_2.6H_2O$, is 226.9, obtained by adding the RAMs of each of the atoms making up the compound. Here, a chemical balance is used to measure accurately 0.1 mole of the substance, which has a mass of 22.69 grams.

The balance has been tared, or set to zero, with the empty beaker on the pan, so that the mass of the sample is displayed

50 ml beaker

Pan

Cobalt chloride is a red solid at room temperature

Accurate chemical balance

Digital readout shows that the mass of the sample is 22.69 grams

The periodic table

THE CHEMICAL ELEMENTS CAN BE arranged according to their **atomic number** (the number of **protons** in the **nuclei** of their **atoms**) and the way in which their **electrons** are organized. The result is the periodic table. **Elements** at the beginning of each horizontal row, or period, have one electron in the outer **electron shell** of their atoms (see pp. 10-11). All of the elements in each vertical column, or group, of the table have similar chemical properties because they all have the same number of outer electrons. The elements of the last group of the table, group 18, have full outer electron shells, and are inert, or unreactive. These elements are called the **noble gases**. Moving down the table, the length of the periods increases in steps, because as the atoms become larger, more types of electron **orbitals** become available. Periods six and seven are 32 elements long, but for simplicity a series of elements from each of these periods is placed separately under the main table.

Group 1	Group 2	Group 3	Group 4	Group 5	Group 6	Group 7	Group 8	Group 9
1 **H** Hydrogen 1.0								
3 **Li** Lithium 6.9	4 **Be** Beryllium 9.0							
11 **Na** Sodium 23.0	12 **Mg** Magnesium 24.3							
19 **K** Potassium 39.1	20 **Ca** Calcium 40.1	21 **Sc** Scandium 45.0	22 **Ti** Titanium 47.9	23 **V** Vanadium 50.9	24 **Cr** Chromium 52.0	25 **Mn** Manganese 54.9	26 **Fe** Iron 55.9	27 **Co** Cobalt 58.9
37 **Rb** Rubidium 85.5	38 **Sr** Strontium 87.6	39 **Y** Yttrium 88.9	40 **Zr** Zirconium 91.2	41 **Nb** Niobium 92.9	42 **Mo** Molybdenum 95.9	43 **Tc** Technetium (99)	44 **Ru** Ruthenium 101.0	45 **Rh** Rhodium 102.9
55 **Cs** Caesium 132.9	56 **Ba** Barium 137.3	57-71	72 **Hf** Hafnium 178.5	73 **Ta** Tantalum 180.9	74 **W** Tungsten 183.9	75 **Re** Rhenium 186.2	76 **Os** Osmium 190.2	77 **Ir** Iridium 192.2
87 **Fr** Francium 223.0	88 **Ra** Radium 226.0	89-103	104 **Unq** Unnilquadium (261)	105 **Unp** Unnilpentium (262)	106 **Unh** Unnilhexium (263)	107 **Uns** Unnilseptium (262)	108 **Uno** Unniloctium (265)	109 **Une** Unnilennium (266)

s-block

Relative atomic mass is estimated, as element exists fleetingly

d-block

Disputes over the discovery and naming of elements 104-109 have led to temporary systematic Latin names

KEY TO TYPES OF ELEMENT

- ALKALI METALS
- ALKALINE EARTH METALS
- TRANSITION METALS
- LANTHANIDES (RARE EARTHS)
- ACTINIDES
- POOR METALS
- SEMI-METALS
- NON-METALS
- NOBLE GASES

57 **La** Lanthanum 138.9	58 **Ce** Cerium 140.1	59 **Pr** Praseodymium 140.9	60 **Nd** Neodymium 144.2	61 **Pm** Promethium (145)	62 **Sm** Samarium 150.4
89 **Ac** Actinium 227.0	90 **Th** Thorium 232.0	91 **Pa** Protactinium 231.0	92 **U** Uranium 238.0	93 **Np** Neptunium (237)	94 **Pu** Plutonium (242)

f-block

Uranium, atomic number 92, is the heaviest element found on Earth. Heavier elements are inherently unstable, because the nuclei of their atoms are too large to hold together. The transuranic elements, atomic numbers 93 to 109, are only produced artificially in the laboratory.

NOBLE GASES

Group 18, on the right of the table, contains elements whose atoms have filled outer electron shells. This means that they are inert elements, reacting with other substances only under extreme conditions, and so forming few **compounds**.

Group 18

| | | | | | | *Group 18* |

Atomic number ——— 5

Chemical symbol ——— B

Name of element ——— Boron

Relative atomic mass ——— 10.8

Group 13	*Group 14*	*Group 15*	*Group 16*	*Group 17*	
					2 **He** Helium 4.0
5 **B** Boron 10.8	6 **C** Carbon 12.0	7 **N** Nitrogen 14.0	8 **O** Oxygen 16.0	9 **F** Fluorine 19.0	10 **Ne** Neon 20.2
13 **Al** Aluminium 27.0	14 **Si** Silicon 28.1	15 **P** Phosphorus 31.0	16 **S** Sulphur 32.1	17 **Cl** Chlorine 35.5	18 **Ar** Argon 40.0

Group 10	*Group 11*	*Group 12*	*Group 13*	*Group 14*	*Group 15*	*Group 16*	*Group 17*	
28 **Ni** Nickel 58.7	29 **Cu** Copper 63.5	30 **Zn** Zinc 65.4	31 **Ga** Gallium 69.7	32 **Ge** Germanium 72.6	33 **As** Arsenic 74.9	34 **Se** Selenium 79.0	35 **Br** Bromine 79.9	36 **Kr** Krypton 83.8
46 **Pd** Palladium 106.4	47 **Ag** Silver 107.9	48 **Cd** Cadmium 112.4	49 **In** Indium 114.8	50 **Sn** Tin 118.7	51 **Sb** Antimony 121.8	52 **Te** Tellurium 127.6	53 **I** Iodine 126.9	54 **Xe** Xenon 131.3
78 **Pt** Platinum 195.1	79 **Au** Gold 197.0	80 **Hg** Mercury 200.6	81 **Tl** Thallium 204.4	82 **Pb** Lead 207.2	83 **Bi** Bismuth 209.0	84 **Po** Polonium 210.0	85 **At** Astatine (211)	86 **Rn** Radon 222.0

d-block

p-block

Different blocks of the periodic table contain elements whose atoms have different orbitals in their outer electron shells

Moving to the adjacent element along a period, atomic number increases by one

Lanthanides and actinides placed separately from rest of periods six and seven

63 **Eu** Europium 152.0	64 **Gd** Gadolinium 157.3	65 **Tb** Terbium 158.9	66 **Dy** Dysprosium 162.5	67 **Ho** Holmium 164.9	68 **Er** Erbium 167.3	69 **Tm** Thulium 168.9	70 **Yb** Ytterbium 173.0	71 **Lu** Lutetium 175.0
95 **Am** Americium (243)	96 **Cm** Curium (247)	97 **Bk** Berkelium (247)	98 **Cf** Californium (251)	99 **Es** Einsteinium (254)	100 **Fm** Fermium (253)	101 **Md** Mendelevium (256)	102 **No** Nobelium (254)	103 **Lr** Lawrencium (257)

f-block

Metals and non-metals

MOST OF THE ELEMENTS ARE METALS. Metals are usually lustrous (shiny), and, apart from copper and gold, are silver or grey in colour. They are all good conductors of heat and electricity, and are ductile (capable of being drawn into wire) and malleable (capable of being hammered into sheets) to different extents. Found at the left-hand side of the periodic table (see pp. 12-13), metals have few outer **electrons**, which they easily lose to form **cations**. Their **compounds** generally exhibit **ionic bonding** (see pp. 16-17). Most non-metals are gases at room temperature, and generally form **anions**. Many simple ionic compounds are formed by metal **atoms** losing electrons to non-metals, and the resulting ions bonding to form **macromolecules**. Sodium and chlorine react in this way to form sodium chloride. In nature, most metals are found not as **elements**, but in compounds known as **ores**. Most metals easily combine with oxygen to form metal oxides, and many ores consist of metal oxides. The simple removal of oxygen is enough to extract a metal from such an ore. The more **reactive** a metal is, the more **energy** is needed for its extraction. Iron can be extracted relatively easily from iron oxide, while more reactive sodium must be extracted by a powerful electric current.

METALLIC ELEMENTS

Like many metals, tin is lustrous

TIN

A layer of grey aluminium oxide coats the particles of aluminium powder

ALUMINIUM POWDER

COPPER TURNINGS

Magnesium, a typical metal, is ductile

MAGNESIUM RIBBON

CATIONS AND ANIONS

Outer electron orbitals

Non-metal atoms have a nearly filled outer electron shell

NON-METAL ATOM

Negative ion

Gaining electrons gives a stable configuration

NON-METAL ANION

Metals have few electrons in their outer shell

Outer electron orbitals

METAL ATOM

Positive ion

Losing outer electrons makes the electron configuration more stable

METAL CATION

FORMATION OF SODIUM CHLORIDE FROM ITS ELEMENTS

When metallic sodium, Na, is gently heated and placed in the non-metallic gas chlorine, Cl_2, a violent **exothermic reaction** occurs. The **product** of the reaction is sodium chloride, NaCl – the familiar white **crystals** of common salt.

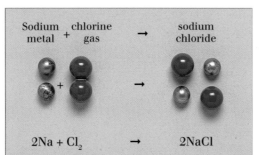

Sodium metal	+	chlorine gas	→	sodium chloride

$$2Na + Cl_2 \rightarrow 2NaCl$$

MOLECULAR VIEW
Each chlorine **molecule** has two chlorine atoms. Two sodium atoms react with each chlorine molecule, to form sodium chloride (see p. 17). Electrons are transferred from the sodium atoms to the chlorine atoms.

Gas jar

Chlorine gas

CHLORINE GAS
Chlorine is a greenish-yellow poisonous gas at room temperature. It is in group 17 of the periodic table.

SODIUM METAL
Like most metals, sodium is silver-grey. It is a soft metal, found in group 1 of the periodic table.

Sodium is easily cut, exposing its lustre

Tiny pieces of sodium chloride form smoke in the jar

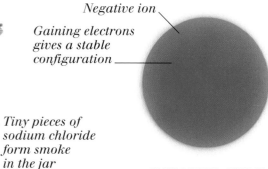

Sodium metal coated with a layer of sodium chloride

Heat of the reaction with chlorine ignites piece of sodium

Sodium chloride

SODIUM CHLORIDE
Sodium chloride is a white solid at room temperature. It consists of macromolecules.

EXTRACTION OF METALS

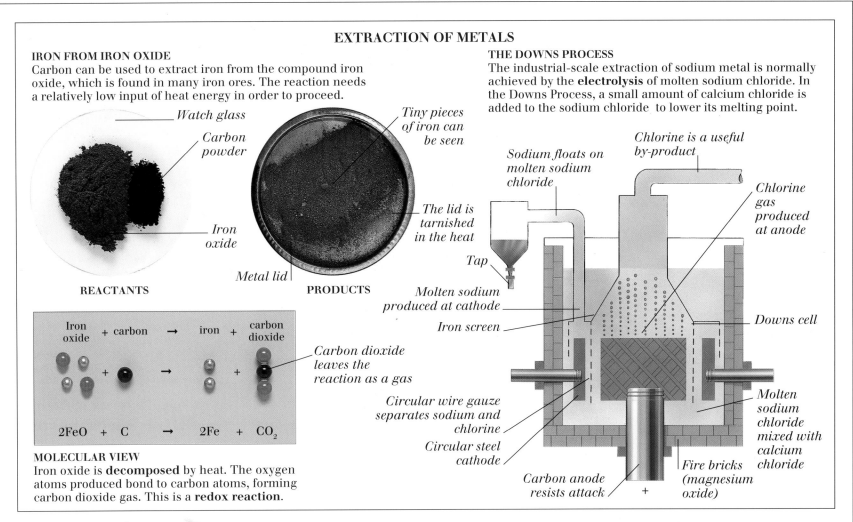

IRON FROM IRON OXIDE
Carbon can be used to extract iron from the compound iron oxide, which is found in many iron ores. The reaction needs a relatively low input of heat energy in order to proceed.

Watch glass

Carbon powder

Iron oxide

REACTANTS

Tiny pieces of iron can be seen

The lid is tarnished in the heat

Metal lid

PRODUCTS

| Iron oxide | + carbon | → | iron | + | carbon dioxide |

$2FeO + C \rightarrow 2Fe + CO_2$

Carbon dioxide leaves the reaction as a gas

MOLECULAR VIEW
Iron oxide is **decomposed** by heat. The oxygen atoms produced bond to carbon atoms, forming carbon dioxide gas. This is a **redox reaction**.

THE DOWNS PROCESS
The industrial-scale extraction of sodium metal is normally achieved by the **electrolysis** of molten sodium chloride. In the Downs Process, a small amount of calcium chloride is added to the sodium chloride to lower its melting point.

Chlorine is a useful by-product

Sodium floats on molten sodium chloride

Chlorine gas produced at anode

Tap

Molten sodium produced at cathode

Iron screen

Downs cell

Circular wire gauze separates sodium and chlorine

Circular steel cathode

Carbon anode resists attack

Molten sodium chloride mixed with calcium chloride

Fire bricks (magnesium oxide)

METALS AND OXYGEN

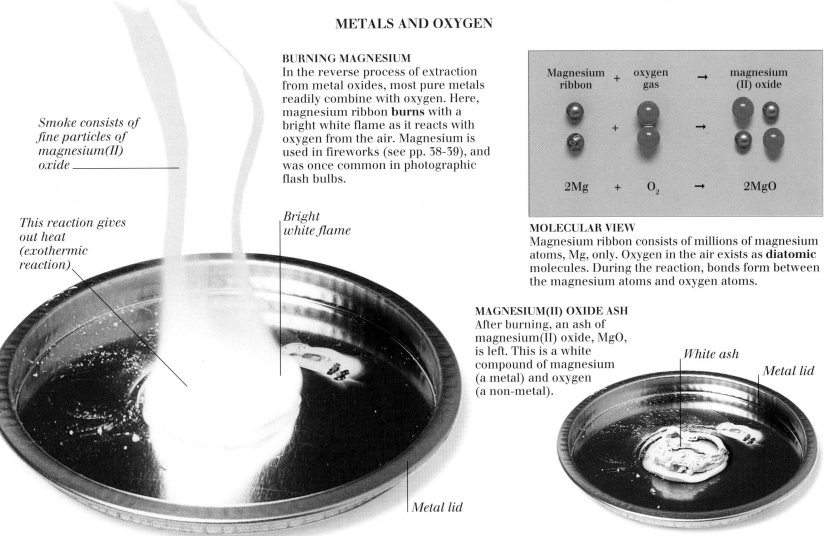

Smoke consists of fine particles of magnesium(II) oxide

This reaction gives out heat (exothermic reaction)

Bright white flame

Metal lid

BURNING MAGNESIUM
In the reverse process of extraction from metal oxides, most pure metals readily combine with oxygen. Here, magnesium ribbon **burns** with a bright white flame as it reacts with oxygen from the air. Magnesium is used in fireworks (see pp. 38-39), and was once common in photographic flash bulbs.

| Magnesium ribbon | + | oxygen gas | → | magnesium (II) oxide |

$2Mg + O_2 \rightarrow 2MgO$

MOLECULAR VIEW
Magnesium ribbon consists of millions of magnesium atoms, Mg, only. Oxygen in the air exists as **diatomic** molecules. During the reaction, bonds form between the magnesium atoms and oxygen atoms.

MAGNESIUM(II) OXIDE ASH
After burning, an ash of magnesium(II) oxide, MgO, is left. This is a white compound of magnesium (a metal) and oxygen (a non-metal).

White ash

Metal lid

Bonds between atoms

ATOMS CAN JOIN – OR BOND – in many ways. Instruments called **atomic force microscopes** produce images of actual **atoms**, revealing these bonds. The two most important types of bonding are **ionic bonding** and **covalent bonding**. **Compounds** are referred to as ionic or covalent depending on the type of bonding that they exhibit. In ionic bonding, a transfer of **electrons** from one atom to another creates two **ions** with opposing **electric charge**. The transfer is generally from a metal to a non-metal (see pp. 14-15). Electrostatic attraction between the ions of opposite charge holds them together. Ionic compounds form **macromolecules** – giant structures consisting of millions of ions. A familiar example of an ionic compound is sodium chloride (common salt). Each grain of common salt is a macromolecule. Atoms that are bound covalently share electrons in their outer **electron shells**. These shared electrons are found within regions called **molecular orbitals**. Another important type of bonding, **hydrogen bonding**, occurs between **molecules** of many hydrogen-containing compounds, and is the cause of some of the unusual properties of water.

This image shows atoms of gold on a graphite surface. The colours are added to the image for clarity. The graphite atoms are joined by covalent bonds.

COVALENT AND IONIC COMPOUNDS

Flame

Gas flame

Glass

Gas burner

The mantle reaches such a high temperature in the gas flame that it glows, but still does not melt

Valve

GAS LAMP, WITH MANTLE

Candle

Wax melts and then vaporizes in the heat of the candle flame

Candle wax is quite soft, and melts easily, like many covalent compounds

RELATIVE MELTING POINTS
A covalent compound melts when the weak bonds between its molecules break. An ionic substance consists of ions held together by strong bonds in a giant macromolecule. More **energy** is needed to break these bonds, so ionic substances generally have higher melting points than covalent ones. Candle wax (covalent) melts at a lower temperature than a gas mantle (ionic), which can be heated until it glows white hot without melting.

CANDLE WAX, A COVALENT COMPOUND

AN EXAMPLE OF IONIC BONDING

1. NEUTRAL ATOMS OF LITHIUM AND FLUORINE

1s-orbital

2s-orbital holds only one electron

2p-orbital

2s-orbital

Atom of lithium, a metallic element

An atom of fluorine, a non-metallic element

Second electron shell holds seven electrons

2. ELECTRON TRANSFER

Shell now holds eight electrons and is filled

Electron transfer

1s-orbital

Lithium atom loses 2s-electron to become a lithium cation, Li$^+$

3. IONIC BONDING: LITHIUM FLUORIDE

Oppositely charged ions attract each other

Li$^+$ ion

Fluorine atom gains an electron to become a fluoride anion, F$^-$

F$^-$ ion

MOLECULAR ORBITALS

The outer electron **orbitals** (see pp. 10-11) of atoms can overlap to form molecular orbitals, which make the covalent bond. Sometimes, s- and p-orbitals of an atom form combined orbitals, called **hybrid orbitals**, prior to forming molecular orbitals.

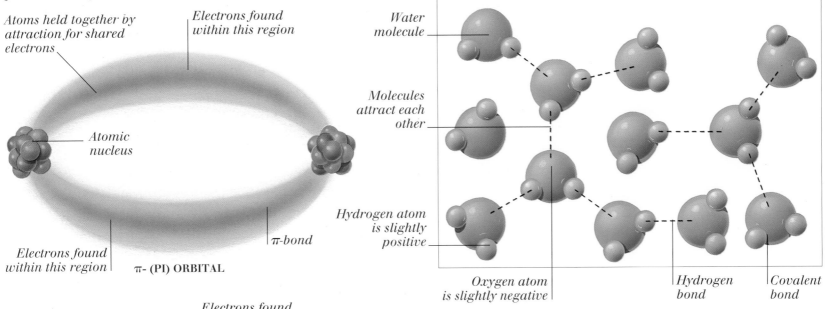

Atoms held together by attraction for shared electrons

Electrons found within this region

Atomic nucleus

Electrons found within this region

π-bond

π- (PI) ORBITAL

σ-bond

Electrons found within this region

Atomic nucleus

σ- (SIGMA) ORBITAL

Electrons found within this region

One s-orbital and three p-orbitals form four sp³ hybrid orbitals

Each sp³ hybrid orbital has this asymmetric dumb-bell shape

Atomic nucleus

Electrons found within this region

SP³ HYBRID ORBITAL

HYDROGEN BONDING

Hydrogen bonds occur between some hydrogen-containing molecules, such as water. In water molecules, negatively charged electrons are concentrated around the oxygen atom, making it slightly negatively charged relative to the hydrogen atoms. Oppositely charged parts of neighbouring molecules attract each other, forming hydrogen bonds.

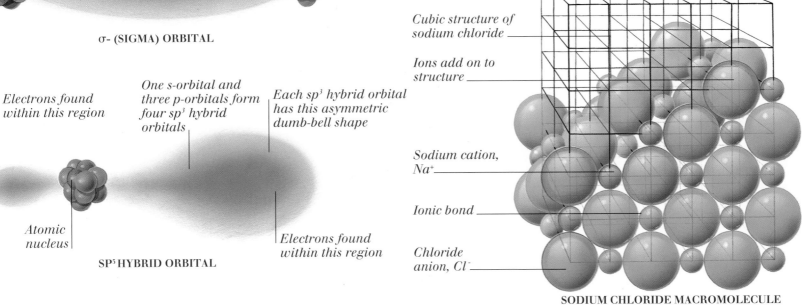

Water molecule

Molecules attract each other

Hydrogen atom is slightly positive

Oxygen atom is slightly negative

Hydrogen bond

Covalent bond

SODIUM CHLORIDE

A macromolecule of sodium chloride forms when sodium **cations** and chloride **anions** bond together. Ions are arranged in the macromolecule in a regular pattern, forming a **crystal**.

Cubic structure of sodium chloride

Ions add on to structure

Sodium cation, Na^+

Ionic bond

Chloride anion, Cl^-

SODIUM CHLORIDE MACROMOLECULE

AN EXAMPLE OF COVALENT BONDING

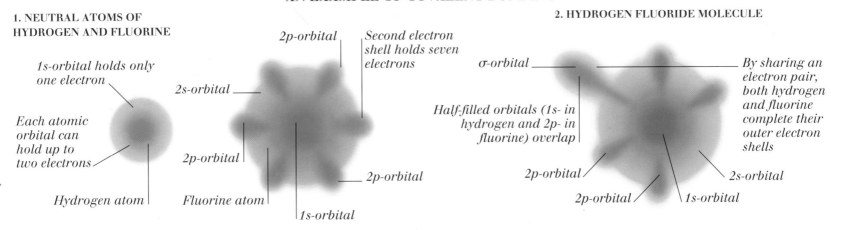

1. NEUTRAL ATOMS OF HYDROGEN AND FLUORINE

1s-orbital holds only one electron

Each atomic orbital can hold up to two electrons

Hydrogen atom

2p-orbital

2s-orbital

2p-orbital

Fluorine atom

1s-orbital

Second electron shell holds seven electrons

2. HYDROGEN FLUORIDE MOLECULE

σ-orbital

Half-filled orbitals (1s- in hydrogen and 2p- in fluorine) overlap

2p-orbital

2p-orbital

By sharing an electron pair, both hydrogen and fluorine complete their outer electron shells

2s-orbital

1s-orbital

Chemical reactions

IN A CHEMICAL REACTION, THE ATOMS or **ions** of the **reactants** are rearranged to give **products** with different chemical and physical properties. For example, **solutions** of lead nitrate and potassium iodide react to produce a solid **precipitate**. Many reactions are **reversible**. Brown nitrogen dioxide gas **decomposes** at high **temperatures** to form a colourless **mixture** of oxygen and nitrogen monoxide. As the mixture cools, nitrogen dioxide forms again. The reactants and products are said to be in an **equilibrium**, the position of which depends on the temperature. Reactant and product **concentrations** may also affect the equilibrium. **Reaction rates** depend upon a number of factors, including temperature and concentration. Marble and dilute **acid** react together more rapidly if the marble is powdered to give it a greater surface area. During a **chemical reaction**, matter is neither created nor destroyed, only changed from one form to another – so the total mass of the products always equals the mass of the reactants.

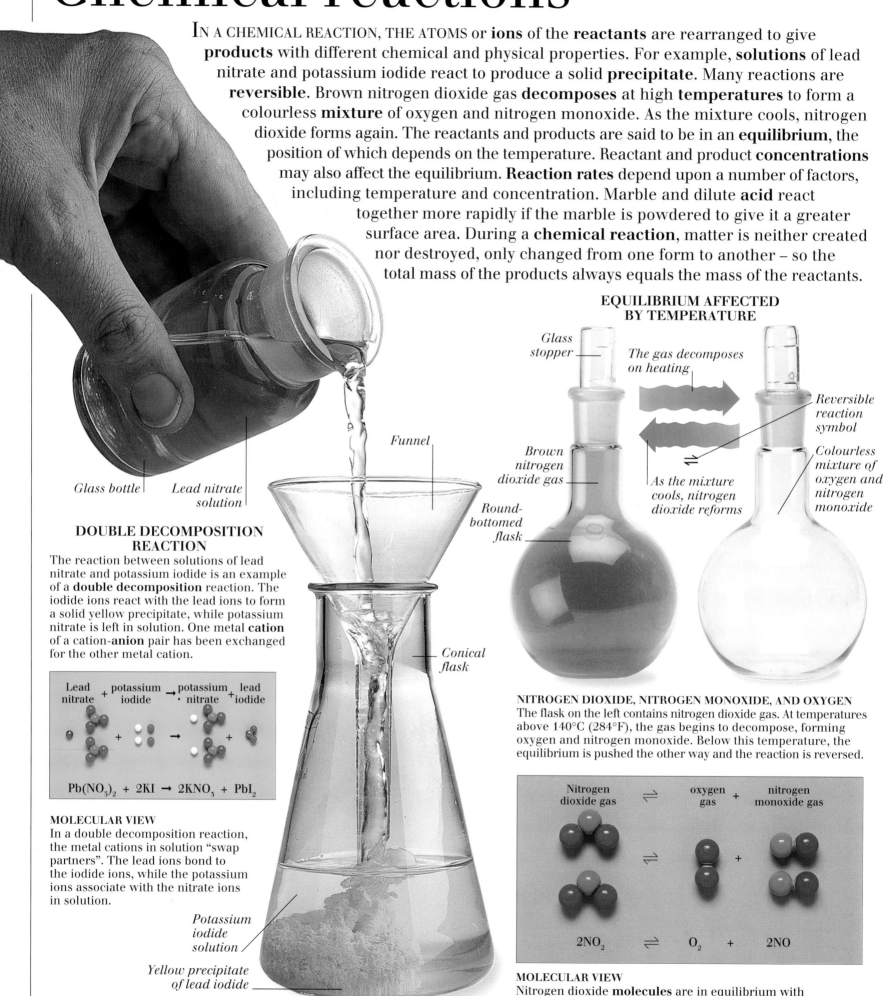

Glass bottle

Lead nitrate
solution

Funnel

*Conical
flask*

*Potassium
iodide
solution*

*Yellow precipitate
of lead iodide*

EQUILIBRIUM AFFECTED BY TEMPERATURE

Glass
stopper

*The gas decomposes
on heating*

*Reversible
reaction
symbol*

Brown
nitrogen
dioxide gas

*Colourless
mixture of
oxygen and
nitrogen
monoxide*

*As the mixture
cools, nitrogen
dioxide reforms*

Round-
bottomed
flask

DOUBLE DECOMPOSITION REACTION
The reaction between solutions of lead nitrate and potassium iodide is an example of a **double decomposition** reaction. The iodide ions react with the lead ions to form a solid yellow precipitate, while potassium nitrate is left in solution. One metal **cation** of a cation-**anion** pair has been exchanged for the other metal cation.

Lead nitrate + potassium iodide → potassium nitrate + lead iodide

$$Pb(NO_5)_2 + 2KI \rightarrow 2KNO_5 + PbI_2$$

MOLECULAR VIEW
In a double decomposition reaction, the metal cations in solution "swap partners". The lead ions bond to the iodide ions, while the potassium ions associate with the nitrate ions in solution.

NITROGEN DIOXIDE, NITROGEN MONOXIDE, AND OXYGEN
The flask on the left contains nitrogen dioxide gas. At temperatures above 140°C (284°F), the gas begins to decompose, forming oxygen and nitrogen monoxide. Below this temperature, the equilibrium is pushed the other way and the reaction is reversed.

Nitrogen dioxide gas ⇌ oxygen gas + nitrogen monoxide gas

$$2NO_2 \rightleftharpoons O_2 + 2NO$$

MOLECULAR VIEW
Nitrogen dioxide **molecules** are in equilibrium with **diatomic** molecules of oxygen and nitrogen monoxide.

EQUILIBRIUM AFFECTED BY CONCENTRATION

COBALT AND CHLORIDE IONS

A pink solution of a cobalt(II) **salt** contains cobalt ions, Co^{2+}. When concentrated hydrochloric acid is added to the solution, chloride ions, Cl^-, cluster around the cobalt ions, forming a **complex ion**, $CoCl_4^{2-}$, in a reversible reaction. The presence of this ion gives the solution a blue colour. Adding more acid pushes the equilibrium position over towards the product – the complex ion. If the concentration of chloride ions is reduced by adding water, the pink colour returns. The addition of water pushes the equilibrium position back towards the reactants – the simple cobalt(II) and chloride ions.

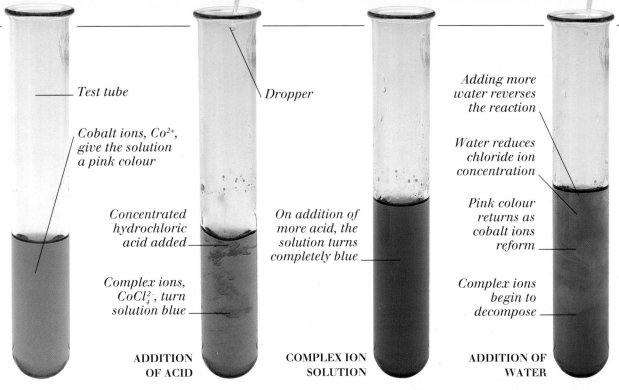

Test tube

Cobalt ions, Co^{2+}, give the solution a pink colour

Concentrated hydrochloric acid added

Complex ions, $CoCl_4^{2-}$, turn solution blue

Dropper

On addition of more acid, the solution turns completely blue

Adding more water reverses the reaction

Water reduces chloride ion concentration

Pink colour returns as cobalt ions reform

Complex ions begin to decompose

COBALT(II) SALT SOLUTION   **ADDITION OF ACID**   **COMPLEX ION SOLUTION**   **ADDITION OF WATER**

RATE OF REACTION

MARBLE CHIPS

Marble is one form of the **ionic** compound, calcium carbonate, $CaCO_3$. Relatively few of the ions making up large chips of marble (below) are found on the chip surfaces – most of the ions are within the chips.

SURFACE AREA OF REACTANT

When dilute sulphuric acid reacts with marble (right), carbon dioxide gas is produced. If powdered marble is used (far right), more ions come into contact with the acid, and the reaction proceeds more rapidly.

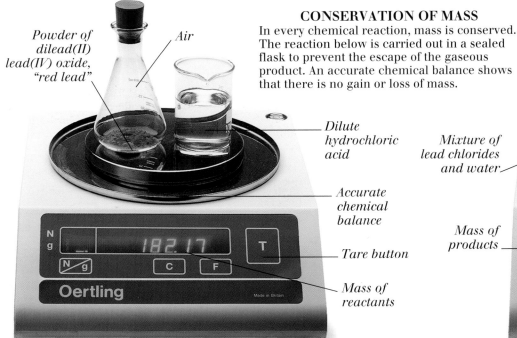

Marble chips

Dilute acid

250 ml beaker

Carbon dioxide gas is produced at a faster rate

Bubbles of carbon dioxide gas are produced slowly

Fine powder of marble

Dilute acid

The mixture fizzes over the beaker

Coarse marble chips

BEAKER WITH CHIPS   **BEAKER WITH POWDER**

CONSERVATION OF MASS

In every chemical reaction, mass is conserved. The reaction below is carried out in a sealed flask to prevent the escape of the gaseous product. An accurate chemical balance shows that there is no gain or loss of mass.

Powder of dilead(II) lead(IV) oxide, "red lead"

Air

Dilute hydrochloric acid

Accurate chemical balance

Tare button

Mass of reactants

Rubber stopper

Chlorine gas and air

Empty beaker

Mas of products

Mixture of lead chlorides and water

Pan

BEFORE THE REACTION
The reactants are weighed before the reaction. The balance is tared (or zeroed) with just the glassware, so that only the mass of the substances inside the glassware will be displayed.

AFTER THE REACTION
The reactants are mixed in the conical flask, and the flask is quickly sealed so that no reaction products can escape. The mass of products is identical to the mass of reactants.

Oxidation and reduction

IN MANY CHEMICAL REACTIONS (see pp. 18-19), **electrons** are transferred between the **atoms** or **ions** taking part. For example, when nitric acid reacts with copper metal, copper atoms lose electrons to become Cu^{2+} ions, while the acid gains electrons. An atom or ion that loses electrons (or gains oxygen) is said to undergo **oxidation**, while an atom or ion that gains electrons (or loses oxygen) undergoes **reduction**. Reactions that involve oxidation and reduction are called **redox reactions**. When an atom or ion is oxidized or reduced, its **oxidation number** changes by the number of electrons transferred. The oxidation number of any atom is 0 (zero), while that of an **element** in a **compound** is given by Roman numerals or by the amount of charge on its ions. For example, iron exists as iron(II) ions, Fe^{2+}, in rust, where it has an oxidation number of +2. An older definition of oxidation was combination with oxygen, as happens in **burning** reactions.

Concentrated nitric acid

Glass tap controls flow of nitric acid into the flask

Separating funnel

Gas is evolved as reaction takes place

Rubber bung

Clamp

Pieces of copper metal

OXIDATION OF COPPER BY NITRIC ACID

A REDOX REACTION
When nitric acid and copper react, each copper atom loses two electrons and is oxidized to copper(II), or Cu^{2+}. Nitric acid, in which nitrogen has an oxidation number of +5, is reduced to nitrogen dioxide, NO_2, also known as nitrogen(IV) oxide, in which nitrogen has an oxidation number of +4.

Glass delivery tube

Gas jar

Round-bottomed flask

Brown nitrogen dioxide gas

Copper +	nitric acid	→	copper nitrate	+	nitrogen dioxide	+ water

$$Cu + 4HNO_3 \rightarrow Cu(NO_3)_2 + 2NO_2 + 2H_2O$$

MOLECULAR MODEL OF REACTION

RUSTING OF IRON
The rusting of iron is an example of a redox reaction. Iron is oxidized to iron(II), with an oxidation number of +2, when it reacts with water and oxygen. The resulting compound, known as rust, is hydrated iron oxide. The tubes below show that both water and oxygen are needed for rust to form.

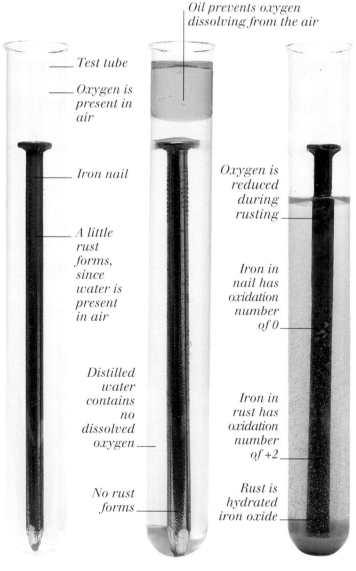

Oil prevents oxygen dissolving from the air

Test tube

Oxygen is present in air

Iron nail

A little rust forms, since water is present in air

Distilled water contains no dissolved oxygen

No rust forms

Oxygen is reduced during rusting

Iron in nail has oxidation number of 0

Iron in rust has oxidation number of +2

Rust is hydrated iron oxide

AIR, NO WATER   **WATER, NO AIR**   **AIR AND WATER**

COMBUSTION REACTION

All combustion reactions are redox reactions. Combustion, or burning, is defined as the rapid **exothermic** combination of a substance with oxygen. Candle wax is a **mixture** of **hydrocarbons**, mainly the **alkane** $C_{18}H_{38}$. Oxygen combines with the carbon atoms present to form carbon dioxide, and with the hydrogen atoms to form water.

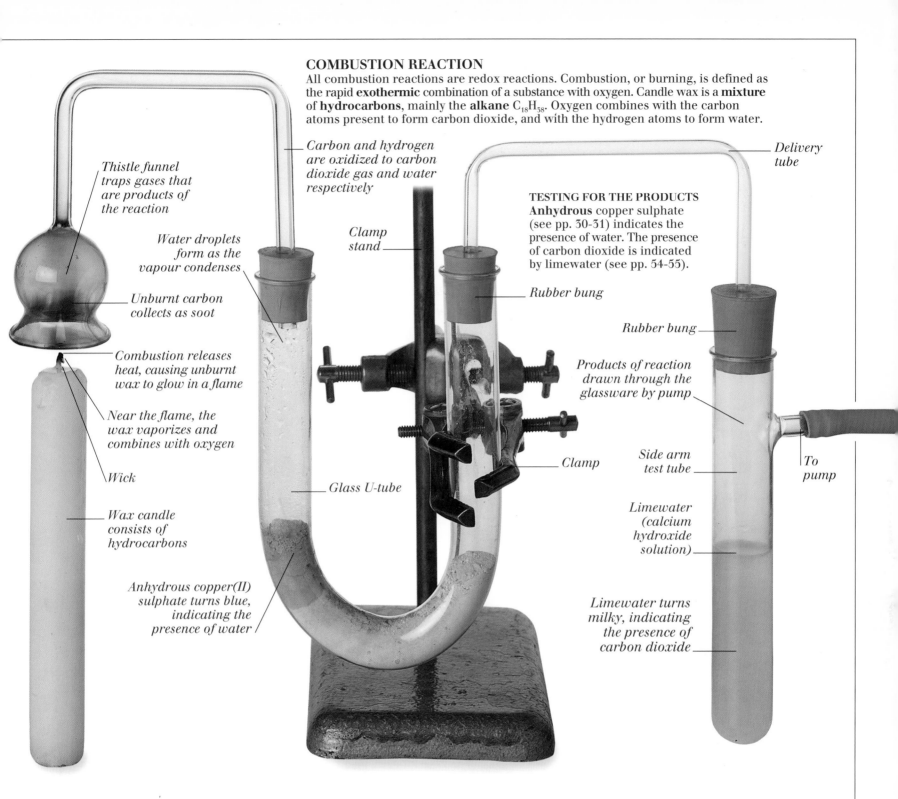

Thistle funnel traps gases that are products of the reaction

Carbon and hydrogen are oxidized to carbon dioxide gas and water respectively

Delivery tube

Water droplets form as the vapour condenses

Clamp stand

TESTING FOR THE PRODUCTS
Anhydrous copper sulphate (see pp. 30-31) indicates the presence of water. The presence of carbon dioxide is indicated by limewater (see pp. 54-55).

Unburnt carbon collects as soot

Rubber bung

Rubber bung

Combustion releases heat, causing unburnt wax to glow in a flame

Products of reaction drawn through the glassware by pump

Near the flame, the wax vaporizes and combines with oxygen

Wick

Glass U-tube

Clamp

Side arm test tube

To pump

Wax candle consists of hydrocarbons

Limewater (calcium hydroxide solution)

Anhydrous copper(II) sulphate turns blue, indicating the presence of water

Limewater turns milky, indicating the presence of carbon dioxide

OXIDATION AS TRANSFER OF ELECTRONS

In many redox reactions, electrons are physically transferred from one atom to another, as shown.

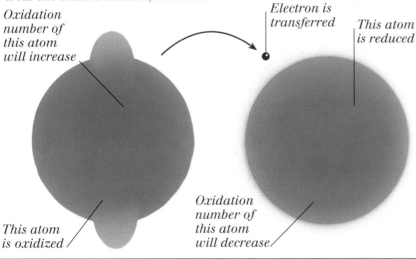

Oxidation number of this atom will increase

Electron is transferred

This atom is reduced

This atom is oxidized

Oxidation number of this atom will decrease

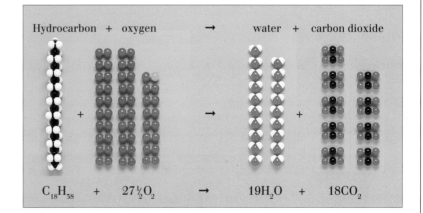

| Hydrocarbon | + | oxygen | → | water | + | carbon dioxide |

$$C_{18}H_{38} \quad + \quad 27\tfrac{1}{2}O_2 \quad \rightarrow \quad 19H_2O \quad + \quad 18CO_2$$

MOLECULAR MODEL OF REACTION
Two **molecules** of the hydrocarbon $C_{18}H_{38}$ react with 55 oxygen molecules, producing 38 molecules of water and 36 of carbon dioxide. Half of these amounts have been shown above.

Acids and bases

ACID IS A COMMON WORD in everyday use, but it has a precise definition in chemistry. An **acid** is defined as a **molecule** or an **ion** that can donate **protons**, or hydrogen ions, H⁺. A **base** is a substance, often an oxide or hydroxide, that accepts protons, and an **alkali** is a base that is water-soluble. Some substances, such as water, can act as either acids or bases, depending on the other substances present. Acids and bases undergo characteristic reactions together, usually in **aqueous solution**, producing a **salt** (see pp. 24-25) and water. In **solution**, acid-base reactions involve the transfer of **hydroxonium ions** or hydrated protons, H_3O^+. These ions form, for example, when hydrogen chloride gas dissolves in water. The **pH scale** gives the **concentration** of hydroxonium ions in solution. As pH falls below 7, a solution becomes more acidic. Conversely, as pH rises above 7, the solution becomes more alkaline. The pH of a solution can be estimated using pigments called **indicators**, or measured accurately with a pH meter.

UNIVERSAL INDICATOR PAPER

This sample of hydrochloric acid has a pH of about 1

This ammonia-based domestic cleaner has a pH of about 10

Strip of universal indicator paper

HYDROCHLORIC ACID

DOMESTIC CLEANER

Pure distilled water has a pH of 7 and is neutral

The pH of liquid soap, a weak alkali, is about 8

Watch glass

DISTILLED WATER

LIQUID SOAP

Digital pH meter

Meter reads pH of 5.83

Knobs to adjust sensitivity

Wire to meter

100 ml beaker

Electronic probe measures concentration of H_3O^+ ions

Bottle of test solution

MEASURING pH
This digital pH meter accurately measures hydroxonium ion concentration. Such meters are often used to find the pH of coloured solutions, which could mask the true colour of indicators.

THE MEANING OF pH

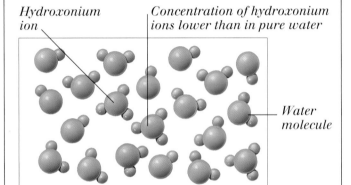

Hydroxonium ion, H_3O^+

Hydroxide ion, OH^-

Water molecule

PURE WATER (NEUTRAL)
Some of the molecules of liquid water break up, or dissociate, forming hydroxide ions, OH^-, and hydrogen ions, H⁺, that become hydrated, H_3O^+. In one litre of pure water at 20°C, there are 10^{-7} **moles** (see pp. 10-11) of each type of ion. This gives a pH value of 7 (neutral) for pure water.

Hydroxonium ion

Concentration of hydroxonium ions lower than in pure water

Water molecule

ALKALINE SOLUTION
When an alkali is added to water, it removes protons, H⁺, from some of the hydroxonium ions, H_3O^+, present, forming more water molecules. The lower the concentration of H_3O^+, the higher the pH. Typically, a weakly alkaline solution has a pH of 10, and a strongly alkaline solution has a pH of 14.

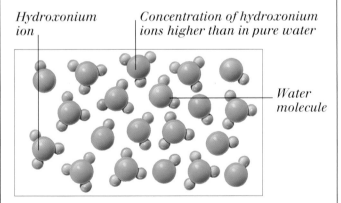

Hydroxonium ion

Concentration of hydroxonium ions higher than in pure water

Water molecule

ACIDIC SOLUTION
When an acid is dissolved in water, it donates protons, H⁺, to water molecules, H_2O, making more hydroxonium ions, H_3O^+. Water thus acts as a base. The concentration of hydroxonium ions increases, and the pH decreases.

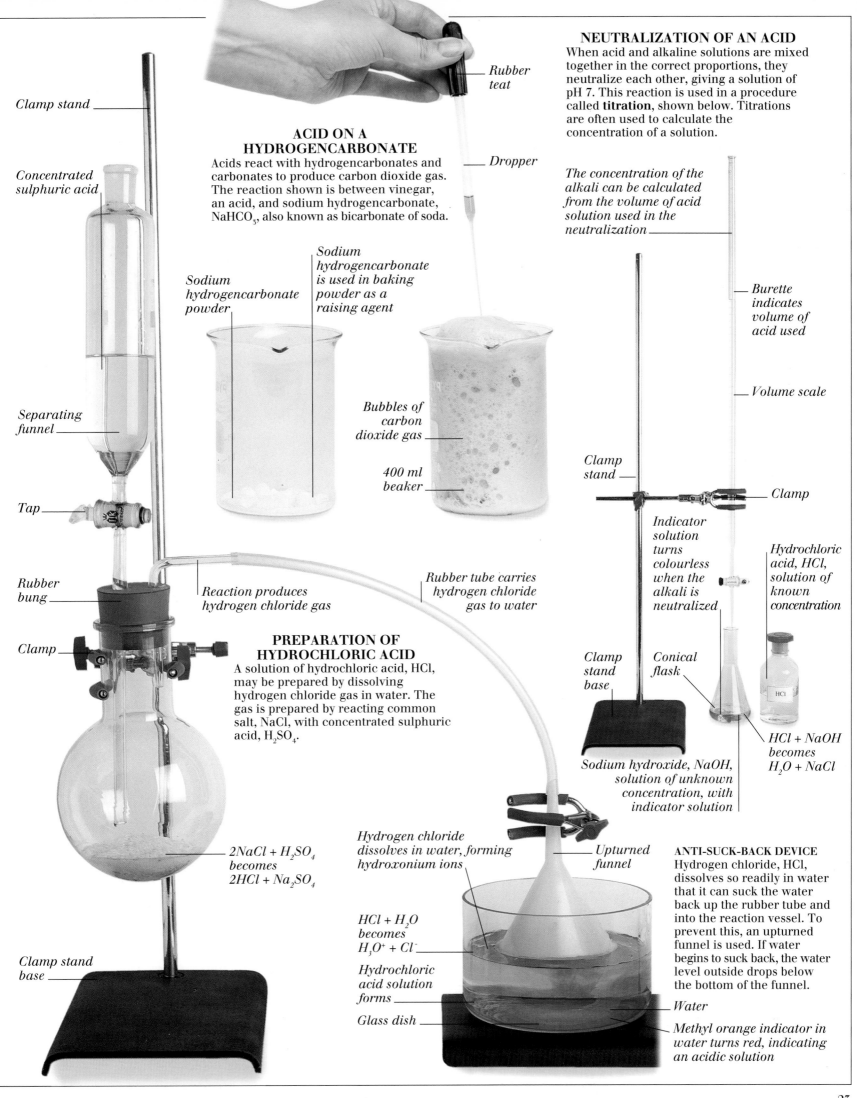

Clamp stand

Concentrated sulphuric acid

Separating funnel

Tap

Rubber bung

Clamp

Clamp stand base

Rubber teat

ACID ON A HYDROGENCARBONATE
Acids react with hydrogencarbonates and carbonates to produce carbon dioxide gas. The reaction shown is between vinegar, an acid, and sodium hydrogencarbonate, $NaHCO_3$, also known as bicarbonate of soda.

Dropper

Sodium hydrogencarbonate powder

Sodium hydrogencarbonate is used in baking powder as a raising agent

Bubbles of carbon dioxide gas

400 ml beaker

Reaction produces hydrogen chloride gas

Rubber tube carries hydrogen chloride gas to water

PREPARATION OF HYDROCHLORIC ACID
A solution of hydrochloric acid, HCl, may be prepared by dissolving hydrogen chloride gas in water. The gas is prepared by reacting common salt, NaCl, with concentrated sulphuric acid, H_2SO_4.

$2NaCl + H_2SO_4$ becomes $2HCl + Na_2SO_4$

Hydrogen chloride dissolves in water, forming hydroxonium ions

$HCl + H_2O$ becomes $H_3O^+ + Cl^-$

Hydrochloric acid solution forms

Glass dish

Upturned funnel

NEUTRALIZATION OF AN ACID
When acid and alkaline solutions are mixed together in the correct proportions, they neutralize each other, giving a solution of pH 7. This reaction is used in a procedure called **titration**, shown below. Titrations are often used to calculate the concentration of a solution.

The concentration of the alkali can be calculated from the volume of acid solution used in the neutralization

Burette indicates volume of acid used

Volume scale

Clamp stand

Clamp

Indicator solution turns colourless when the alkali is neutralized

Hydrochloric acid, HCl, solution of known concentration

Clamp stand base

Conical flask

HCl + NaOH becomes $H_2O + NaCl$

Sodium hydroxide, NaOH, solution of unknown concentration, with indicator solution

ANTI-SUCK-BACK DEVICE
Hydrogen chloride, HCl, dissolves so readily in water that it can suck the water back up the rubber tube and into the reaction vessel. To prevent this, an upturned funnel is used. If water begins to suck back, the water level outside drops below the bottom of the funnel.

Water

Methyl orange indicator in water turns red, indicating an acidic solution

Salts

WHENEVER AN ACID AND A BASE **neutralize** each other (see pp. 22-23), the **products** of the reaction always include a **salt**. A salt is a **compound** that consists of **cations** (positive **ions**) and **anions** (negative ions). The cation is usually a metal ion, such as the sodium ion, Na$^+$. The anion can be a non-metal such as the chloride ion, Cl$^-$, although more often it is a unit called a **radical**. This is a combination of non-metals that remains unchanged during most reactions. So, for example, when copper(II) oxide is added to sulphuric acid, the sulphate radical (SO$_4^{2-}$) becomes associated with copper ions, forming the salt copper(II) sulphate, CuSO$_4$. Salts are very widespread compounds – the most familiar being sodium chloride, or common salt. Mineral water contains salts, which are formed when slightly **acidic** rainwater dissolves rocks such as limestone. Water that contains large amounts of certain dissolved salts is called hard water (see pp. 38-39). A class of salts called acid salts contains a positive hydrogen ion in addition to the usual metal cation. Acid salts can be prepared by careful **titration** of an acid and a **base**.

FORMATION OF SALTS

In the generalized equations below, an acid reacts with three typical bases – a hydroxide, an oxide, and a carbonate. A cation from the base combines with the acid's anion or negative radical, displacing the hydrogen ion to form a salt.

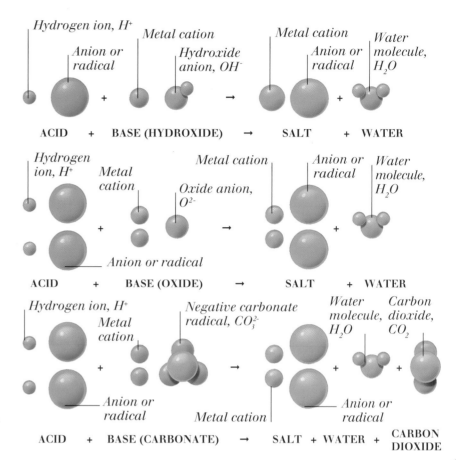

Hydrogen ion, H$^+$ *Metal cation* *Metal cation* *Water molecule, H$_2$O*
Anion or radical *Hydroxide anion, OH$^-$* *Anion or radical*

ACID + BASE (HYDROXIDE) → SALT + WATER

Hydrogen ion, H$^+$ *Metal cation* *Metal cation* *Anion or radical* *Water molecule, H$_2$O*
Oxide anion, O^{2-}
Anion or radical

ACID + BASE (OXIDE) → SALT + WATER

Hydrogen ion, H$^+$ *Metal cation* *Negative carbonate radical, CO$_3^{2-}$* *Water molecule, H$_2$O* *Carbon dioxide, CO$_2$*
Anion or radical *Metal cation* *Anion or radical*

ACID + BASE (CARBONATE) → SALT + WATER + CARBON DIOXIDE

COPPER(II) SULPHATE

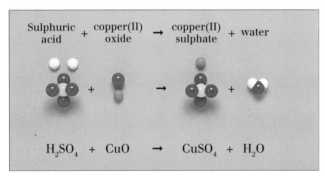

Sulphuric acid + copper(II) oxide → copper(II) sulphate + water

$$H_2SO_4 + CuO \rightarrow CuSO_4 + H_2O$$

MOLECULAR VIEW
When the base copper(II) oxide reacts with sulphuric acid, copper(II) ions take the place of the hydrogen in the acid. The salt formed is therefore copper(II) sulphate. Water is the other product. The sulphate ion is a radical.

COPPER(II) OXIDE AND SULPHURIC ACID
Copper(II) oxide, a black powder, is a base. When added to colourless dilute sulphuric acid, a neutralization reaction occurs. Hydrogen from the acid and oxygen from copper(II) oxide form water, while copper ions and the sulphate radical form the salt copper(II) sulphate.

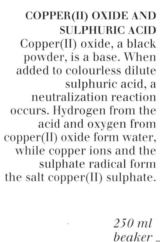

Black copper(II) oxide

Spatula

Salt forms as copper(II) oxide neutralizes acid

250 ml beaker

Dilute sulphuric acid

Blue solution contains copper ions, Cu^{2+}

Copper(II) oxide powder

Watch glass

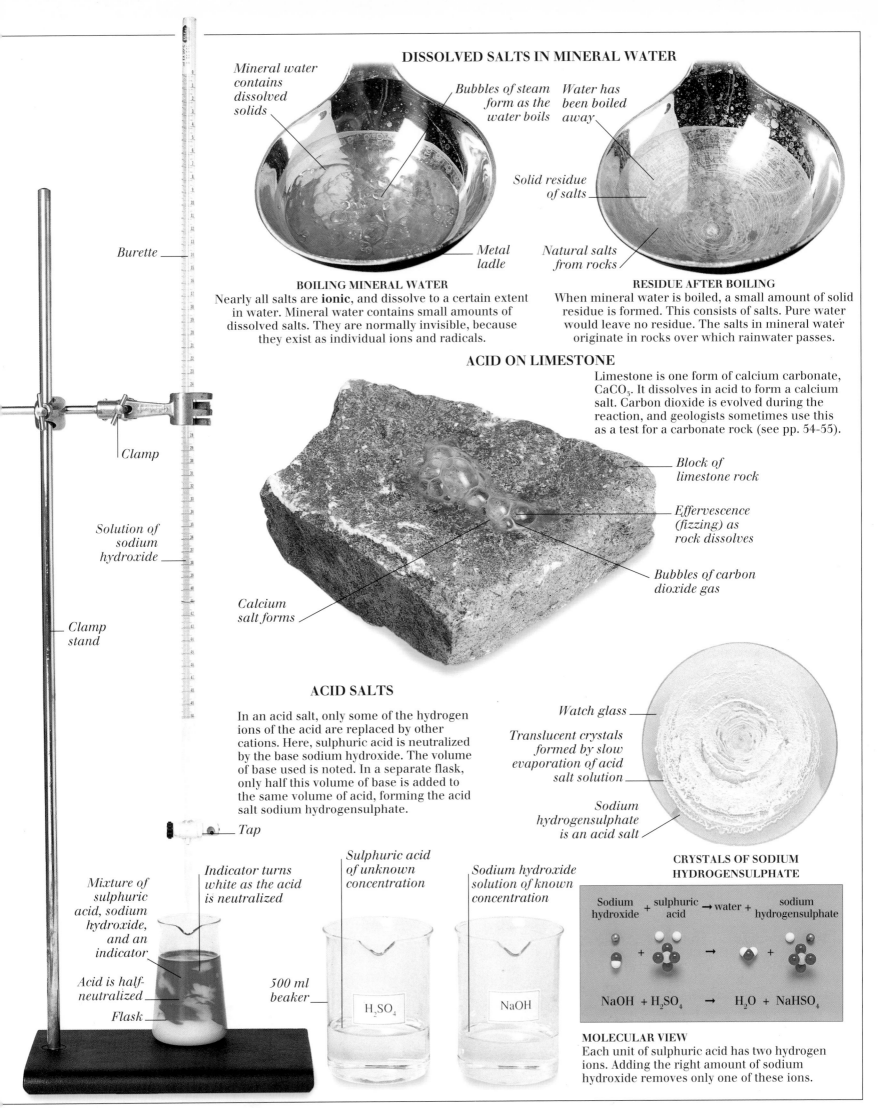

DISSOLVED SALTS IN MINERAL WATER

Mineral water contains dissolved solids

Bubbles of steam form as the water boils

Water has been boiled away

Solid residue of salts

Metal ladle

Natural salts from rocks

BOILING MINERAL WATER
Nearly all salts are **ionic**, and dissolve to a certain extent in water. Mineral water contains small amounts of dissolved salts. They are normally invisible, because they exist as individual ions and radicals.

RESIDUE AFTER BOILING
When mineral water is boiled, a small amount of solid residue is formed. This consists of salts. Pure water would leave no residue. The salts in mineral water originate in rocks over which rainwater passes.

ACID ON LIMESTONE

Limestone is one form of calcium carbonate, $CaCO_3$. It dissolves in acid to form a calcium salt. Carbon dioxide is evolved during the reaction, and geologists sometimes use this as a test for a carbonate rock (see pp. 54-55).

Block of limestone rock

Effervescence (fizzing) as rock dissolves

Bubbles of carbon dioxide gas

Calcium salt forms

Burette

Clamp

Solution of sodium hydroxide

Clamp stand

Tap

ACID SALTS

In an acid salt, only some of the hydrogen ions of the acid are replaced by other cations. Here, sulphuric acid is neutralized by the base sodium hydroxide. The volume of base used is noted. In a separate flask, only half this volume of base is added to the same volume of acid, forming the acid salt sodium hydrogensulphate.

Watch glass

Translucent crystals formed by slow evaporation of acid salt solution

Sodium hydrogensulphate is an acid salt

CRYSTALS OF SODIUM HYDROGENSULPHATE

Mixture of sulphuric acid, sodium hydroxide, and an indicator

Indicator turns white as the acid is neutralized

Acid is half-neutralized

Flask

Sulphuric acid of unknown concentration

Sodium hydroxide solution of known concentration

500 ml beaker

H_2SO_4

NaOH

Sodium hydroxide + sulphuric acid → water + sodium hydrogensulphate

$$NaOH + H_2SO_4 \rightarrow H_2O + NaHSO_4$$

MOLECULAR VIEW
Each unit of sulphuric acid has two hydrogen ions. Adding the right amount of sodium hydroxide removes only one of these ions.

Catalysts

A CATALYST IS A SUBSTANCE that increases the **rate** at which a reaction takes place, but is unchanged itself at the end of the reaction. Certain **catalysts** are used up in one stage of a reaction, and regenerated at a later stage. Light is sometimes considered to be a catalyst – although it is not a substance – because it speeds up certain reactions. This process is referred to as photocatalysis, and is very important in photography and in **photosynthesis** (see pp. 38-39). Often, catalysts simply provide a suitable surface upon which the reaction can take place. Such surface catalysis often involves **transition metals**, such as iron or nickel. Surface catalysis occurs in catalytic converters in motor cars, which speed up reactions that change harmful pollutant gases into less harmful ones. **Enzymes** are biological catalysts and are nearly all **proteins**. They catalyse reactions in living organisms. For example, an enzyme called ptyalin in saliva helps to digest or break down starch in food to make sugars that can be readily absorbed by the body. Enzymes are also important in turning sugar into alcohol during fermentation.

CATALYSIS AT A SURFACE

Reactant A is a diatomic molecule

Reactant B approaches surface

Surface atoms of catalyst

REACTANTS APPROACH SURFACE
In this reaction, one of the **reactants** is a **diatomic molecule** that must be split before it will react.

Atom of diatomic molecule

Reactant bonds weakly to surface atom

REACTANTS BOND TO SURFACE
The reactants form weak bonds with the surface atoms. As the diatomic molecule bonds, it breaks into two individual atoms.

Reactants migrate across surface

Reaction occurs at surface

REACTION TAKES PLACE
The reactants move, or migrate, across the surface. When they meet, the reaction takes place. The surface is unchanged.

Product of reaction

Catalyst surface is unchanged

PRODUCT LEAVES SURFACE
The reaction **product** leaves the surface, to which it was very weakly bonded, and the reaction is complete.

PHOTOCATALYSIS

Light can promote, or speed up, a reaction. Here, both tubes contain a yellow **precipitate** of silver bromide (see pp. 54-55). For a period of about ten minutes, one of the tubes has been left in a dark cupboard while the other has been left in the light. The light has caused silver **ions** to become **atoms** of silver. Photographic films contain tiny granules of silver halides, which produce silver on the negative wherever it is hit by light.

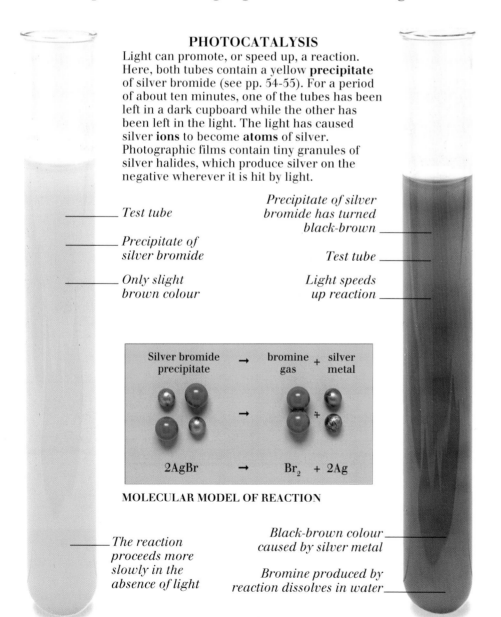

Test tube

Precipitate of silver bromide

Only slight brown colour

Precipitate of silver bromide has turned black-brown

Test tube

Light speeds up reaction

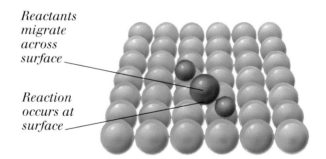

Silver bromide precipitate	→	bromine gas	+	silver metal
2AgBr	→	Br₂	+	2Ag

MOLECULAR MODEL OF REACTION

The reaction proceeds more slowly in the absence of light

Black-brown colour caused by silver metal

Bromine produced by reaction dissolves in water

TUBE LEFT IN DARKNESS

TUBE LEFT IN LIGHT

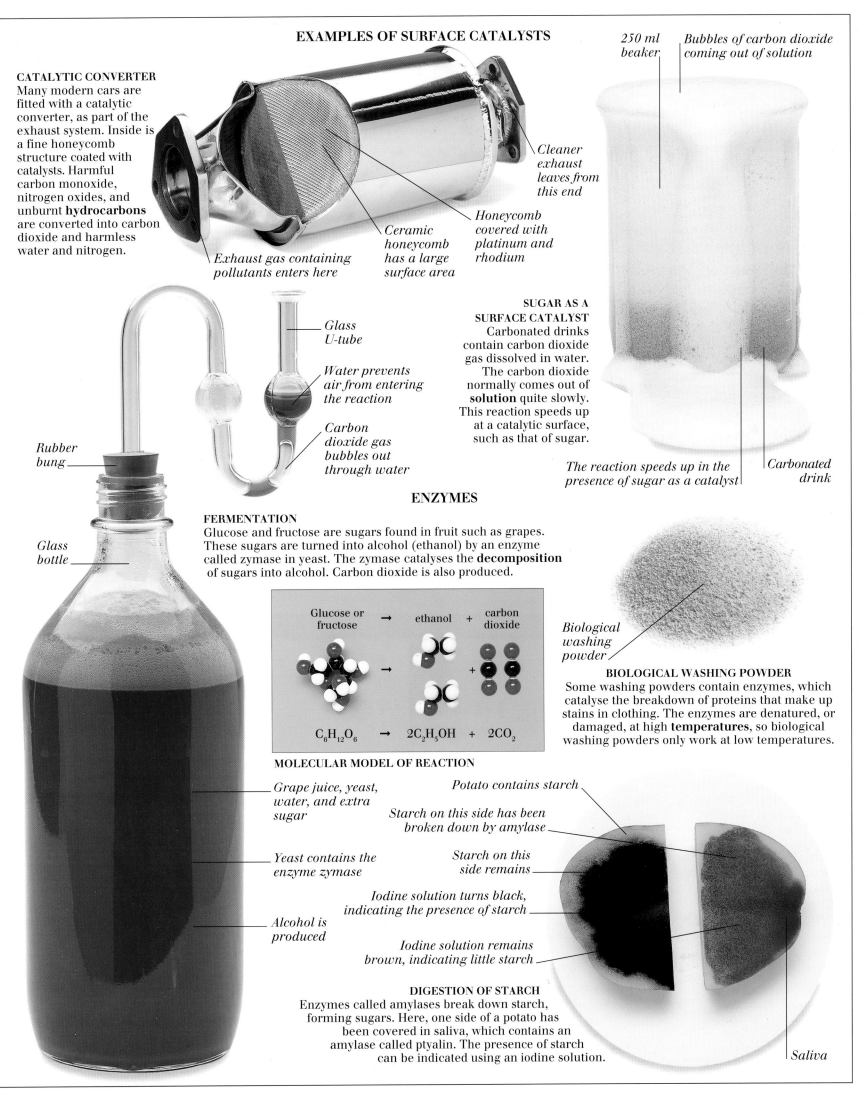

EXAMPLES OF SURFACE CATALYSTS

CATALYTIC CONVERTER
Many modern cars are fitted with a catalytic converter, as part of the exhaust system. Inside is a fine honeycomb structure coated with catalysts. Harmful carbon monoxide, nitrogen oxides, and unburnt **hydrocarbons** are converted into carbon dioxide and harmless water and nitrogen.

250 ml beaker

Bubbles of carbon dioxide coming out of solution

Cleaner exhaust leaves from this end

Honeycomb covered with platinum and rhodium

Ceramic honeycomb has a large surface area

Exhaust gas containing pollutants enters here

Glass U-tube

Water prevents air from entering the reaction

Carbon dioxide gas bubbles out through water

SUGAR AS A SURFACE CATALYST
Carbonated drinks contain carbon dioxide gas dissolved in water. The carbon dioxide normally comes out of **solution** quite slowly. This reaction speeds up at a catalytic surface, such as that of sugar.

The reaction speeds up in the presence of sugar as a catalyst

Carbonated drink

Rubber bung

Glass bottle

ENZYMES

FERMENTATION
Glucose and fructose are sugars found in fruit such as grapes. These sugars are turned into alcohol (ethanol) by an enzyme called zymase in yeast. The zymase catalyses the **decomposition** of sugars into alcohol. Carbon dioxide is also produced.

Glucose or fructose	→	ethanol	+	carbon dioxide

$$C_6H_{12}O_6 \rightarrow 2C_2H_5OH + 2CO_2$$

MOLECULAR MODEL OF REACTION

Biological washing powder

BIOLOGICAL WASHING POWDER
Some washing powders contain enzymes, which catalyse the breakdown of proteins that make up stains in clothing. The enzymes are denatured, or damaged, at high **temperatures**, so biological washing powders only work at low temperatures.

Grape juice, yeast, water, and extra sugar

Yeast contains the enzyme zymase

Alcohol is produced

Potato contains starch

Starch on this side has been broken down by amylase

Starch on this side remains

Iodine solution turns black, indicating the presence of starch

Iodine solution remains brown, indicating little starch

DIGESTION OF STARCH
Enzymes called amylases break down starch, forming sugars. Here, one side of a potato has been covered in saliva, which contains an amylase called ptyalin. The presence of starch can be indicated using an iodine solution.

Saliva

Heat in chemistry

HEAT IS A FORM OF ENERGY that a substance possesses due to the movement or vibration of its **atoms**, **molecules**, or **ions**. The **temperature** of a substance is a measure of the average heat (or **kinetic**) **energy** of its particles, and is a factor in determining whether the substance is solid, liquid, or gas. Energy changes are involved in all reactions. For example, light energy (see pp. 38-39) and electrical energy (see pp. 34-35) can make reactions occur or can be released as a result of reactions. Heat energy is taken in or released by most reactions. Some reactions, such as the **burning** of wood, need an initial input of energy, called **activation energy**, in order for them to occur. Once established, however, the burning reaction releases heat energy to the surroundings – it is an **exothermic** reaction. Other reactions take heat from their surroundings, and are called **endothermic** reactions. The thermite reaction, in which aluminium metal reacts with a metal oxide, is so exothermic that the heat released can be used to weld metals.

LIQUID CHLORINE

A gas becomes a liquid if cooled below its boiling point. Here, chlorine gas has been pumped into a test tube. Heat energy is then removed from the gas by cooling the tube in dry ice.

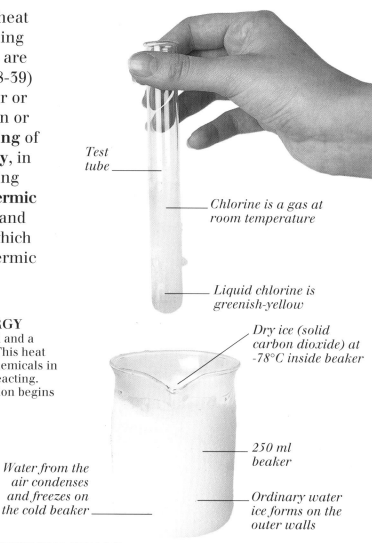

Test tube

Chlorine is a gas at room temperature

Liquid chlorine is greenish-yellow

Dry ice (solid carbon dioxide) at -78°C inside beaker

250 ml beaker

Ordinary water ice forms on the outer walls

ACTIVATION ENERGY

Friction between a match head and a rough surface produces heat. This heat provides the energy that the chemicals in the match head need to start reacting. The heat released in this reaction begins the burning of the wood.

Match rubbed against rough surface

Burning wood combines with oxygen from the air

Match head contains phosphorus

Water from the air condenses and freezes on the cold beaker

Rough surface

EXOTHERMIC AND ENDOTHERMIC REACTIONS

EXOTHERMIC REACTION, $CaCl_2 \rightarrow Ca^{2+} + 2Cl^-$
Compounds contain a certain amount of energy. If the energy of the **products** of a reaction is less than that of the **reactants**, then heat will be released to the surroundings. The reaction is described as exothermic. When calcium chloride dissolves in water, an exothermic reaction takes place.

ENDOTHERMIC REACTION, $NH_4NO_3 \rightarrow NH_4^+ + NO_3^-$
If the energy of the products of a reaction is more than that of the reactants, then heat will be taken from the surroundings. The reaction is described as endothermic. An endothermic reaction occurs when ammonium nitrate is dissolved in water.

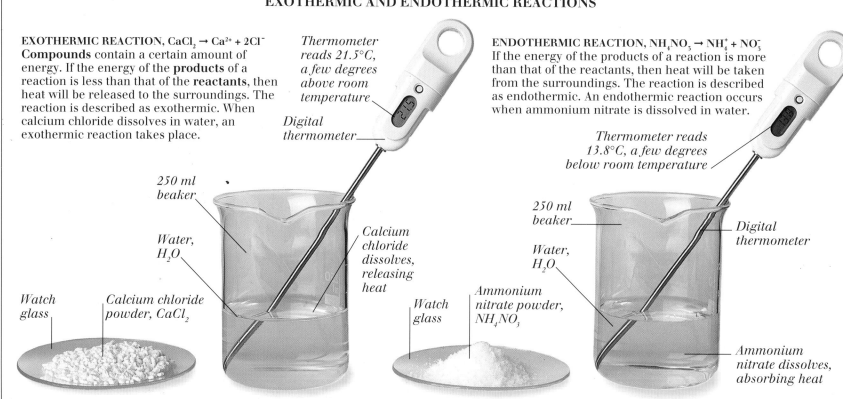

Thermometer reads 21.5°C, a few degrees above room temperature

Digital thermometer

250 ml beaker

Water, H_2O

Calcium chloride dissolves, releasing heat

Watch glass

Calcium chloride powder, $CaCl_2$

Thermometer reads 13.8°C, a few degrees below room temperature

250 ml beaker

Water, H_2O

Digital thermometer

Watch glass

Ammonium nitrate powder, NH_4NO_3

Ammonium nitrate dissolves, absorbing heat

THERMITE REACTION

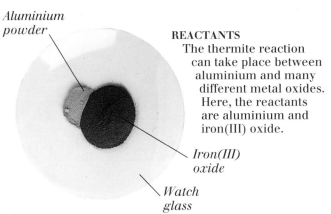

Aluminium powder

Iron(III) oxide

Watch glass

REACTANTS
The thermite reaction can take place between aluminium and many different metal oxides. Here, the reactants are aluminium and iron(III) oxide.

Thick smoke consists of small particles of reaction products

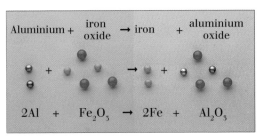

Aluminium +	iron oxide	→ iron	+	aluminium oxide

$$2Al + Fe_2O_3 \rightarrow 2Fe + Al_2O_3$$

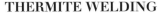

MOLECULAR MODEL OF REACTION

THERMITE WELDING
The tremendous amount of heat released by the thermite reaction is put to good use in welding railway tracks. Iron oxide is used, yielding molten iron as one of the reaction products. The molten iron helps to make the weld.

Pot containing reactants

Molten iron flows into gap to make weld

THE REACTION
When aluminium reacts with iron(III) oxide, aluminium(III) oxide and iron are produced. Aluminium is a very **reactive** metal (see pp. 32-33), and has a greater affinity for oxygen than iron does. The reaction products have much less energy than the reactants, so the reaction of aluminium with iron(III) oxide is exothermic.

Metal tray

A large amount of heat is released

Burning magnesium strip provides the activation energy for the reaction

Products of the reaction are aluminium oxide and metallic iron

Flames

Shower of sparks

Water in chemistry

EACH MOLECULE OF WATER consists of two **atoms** of hydrogen bound to an oxygen atom. Water reacts physically and chemically with a wide range of **elements** and **compounds**. Many gases dissolve in water – in particular, ammonia dissolves very readily, as demonstrated by the fountain experiment. Some compounds, called **dehydrating agents**, have such a strong affinity for water that they can remove it from other substances. Concentrated sulphuric acid is so powerful a dehydrating agent that it can remove hydrogen and oxygen from certain compounds, making water where there was none before. Water is often held in **crystals** of other substances, and is then called **water of crystallization**. A compound can lose its water of crystallization during strong heating, and is then said to be **anhydrous**. Adding water to anhydrous crystals can restore the water of crystallization. Some compounds, described as **efflorescent**, have crystals that lose their water of crystallization to the air. Conversely, **hygroscopic** compounds have crystals that absorb water from the air. Desiccators often employ such compounds to dry other substances.

WATER OF CRYSTALLIZATION
Crystals containing water of crystallization are said to be hydrated. Heating a hydrated crystal causes it to lose water.

Glass dish

Blue solution of copper(II) sulphate

Blue crystals form on evaporation

COPPER(II) SULPHATE SOLUTION
Gently heating a **solution** of blue copper(II) sulphate evaporates the water, leaving behind blue crystals of hydrated copper(II) sulphate.

Strongly heated crystals dehydrate

Gauze

Tripod

Gas flame

Bunsen burner

ANHYDROUS COPPER(II) SULPHATE
Strongly heating the hydrated crystals drives off the water of crystallization, leaving a white powder of anhydrous copper(II) sulphate.

SULPHURIC ACID AS A DEHYDRATING AGENT
Substances known as dehydrating agents can either simply remove water from a **mixture**, or remove hydrogen and oxygen from a compound in the ratio 2:1, the ratio found in water. Concentrated sulphuric acid is a very powerful dehydrating agent (below).

H_2SO_4

CONCENTRATED SULPHURIC ACID, H_2SO_4

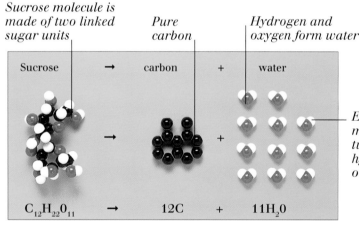

Sucrose molecule is made of two linked sugar units

Pure carbon

Hydrogen and oxygen form water

Sucrose	→	carbon	+	water

Each water molecule has two atoms of hydrogen and one of oxygen

$C_{12}H_{22}O_{11}$ → 12C + 11H$_2$O

MOLECULAR MODEL OF REACTION

Glass dish

Steam condenses on glass

All the hydrogen and oxygen will eventually be removed from the sucrose

Carbon

Sucrose (sugar)

DEHYDRATION OF SUCROSE
Concentrated sulphuric acid removes 22 hydrogen atoms and 11 oxygen atoms from each molecule of sucrose, leaving only black carbon behind. The reaction evolves heat, enough to boil the water produced and form steam.

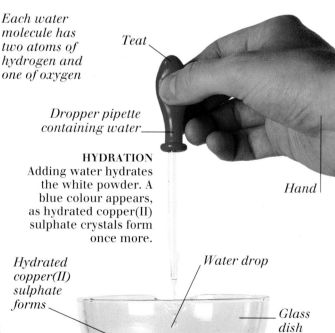

Teat

Dropper pipette containing water

Hand

HYDRATION
Adding water hydrates the white powder. A blue colour appears, as hydrated copper(II) sulphate crystals form once more.

Hydrated copper(II) sulphate forms

Water drop

Glass dish

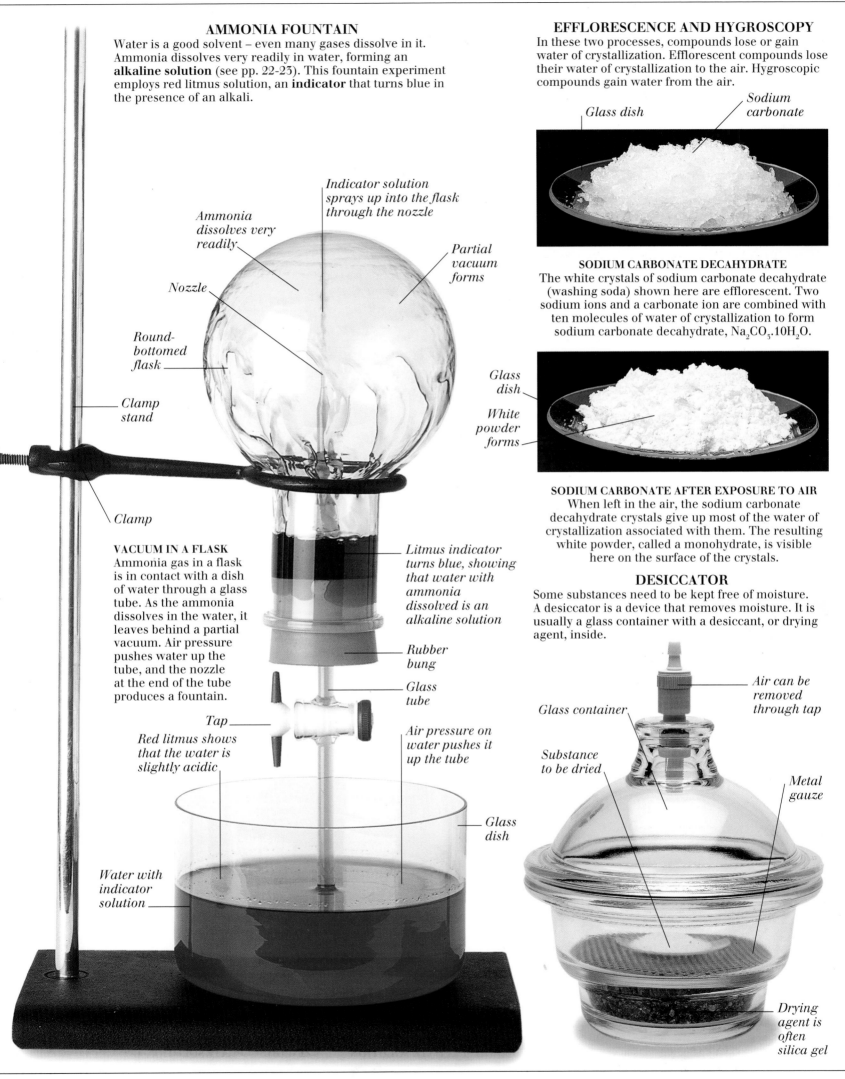

AMMONIA FOUNTAIN

Water is a good solvent – even many gases dissolve in it. Ammonia dissolves very readily in water, forming an **alkaline solution** (see pp. 22-23). This fountain experiment employs red litmus solution, an **indicator** that turns blue in the presence of an alkali.

Indicator solution sprays up into the flask through the nozzle

Ammonia dissolves very readily

Nozzle

Partial vacuum forms

Round-bottomed flask

Clamp stand

Clamp

VACUUM IN A FLASK
Ammonia gas in a flask is in contact with a dish of water through a glass tube. As the ammonia dissolves in the water, it leaves behind a partial vacuum. Air pressure pushes water up the tube, and the nozzle at the end of the tube produces a fountain.

Litmus indicator turns blue, showing that water with ammonia dissolved is an alkaline solution

Rubber bung

Glass tube

Tap

Red litmus shows that the water is slightly acidic

Air pressure on water pushes it up the tube

Glass dish

Water with indicator solution

EFFLORESCENCE AND HYGROSCOPY

In these two processes, compounds lose or gain water of crystallization. Efflorescent compounds lose their water of crystallization to the air. Hygroscopic compounds gain water from the air.

Glass dish

Sodium carbonate

SODIUM CARBONATE DECAHYDRATE
The white crystals of sodium carbonate decahydrate (washing soda) shown here are efflorescent. Two sodium ions and a carbonate ion are combined with ten molecules of water of crystallization to form sodium carbonate decahydrate, $Na_2CO_3.10H_2O$.

Glass dish

White powder forms

SODIUM CARBONATE AFTER EXPOSURE TO AIR
When left in the air, the sodium carbonate decahydrate crystals give up most of the water of crystallization associated with them. The resulting white powder, called a monohydrate, is visible here on the surface of the crystals.

DESICCATOR
Some substances need to be kept free of moisture. A desiccator is a device that removes moisture. It is usually a glass container with a desiccant, or drying agent, inside.

Air can be removed through tap

Glass container

Substance to be dried

Metal gauze

Drying agent is often silica gel

The activity series

ALL METAL ATOMS LOSE ELECTRONS fairly easily and become positive **ions**, or **cations**. The ease with which a metal loses **electrons** is a measure of its **reactivity**. Metals in groups 1 and 2 of the periodic table (see pp. 36-39), which have one and two outer electrons respectively, are usually the most reactive. Aluminium in group 3 is a reactive metal, but less so than calcium in group 2. Metals can be arranged in order of decreasing reactivity in a series known as the activity series. In this series, zinc is placed above copper, and copper above silver. Zinc metal is more reactive than copper and can displace copper ions from a **solution**. Similarly, copper displaces silver from solution. Electrons from the more reactive metal transfer to the less reactive metal ions in solution, resulting in the **deposition** of the less reactive metal. Because electron transfer occurs in these reactions, they are classified as **redox reactions**. The reactivity of a metal may be characterized in many ways – for example, by its reactions with **acids**. The different reactivities of metals have a practical application in the prevention of corrosion in underwater pipes.

TABLE OF METAL REACTIVITY

Metal	Air or oxygen on metal	Water on metal	Acids on metal	Metals on salts of other metals
K Na Ca Mg	Burn in air or oxygen	React with cold water (with decreasing ease)	Displace hydrogen from acids that are not oxidizing agents (with decreasing ease)	Displace a metal lower in the series from a solution of one of its salts
Al Zn Fe		React with steam when heated		
Sn Pb Cu Hg	Converted into the oxide by heating in air	No reaction with water or steam	React only with oxidizing acids	
Ag Au Pt	Unaffected by air or oxygen		No reaction with acids	

ALUMINIUM METAL

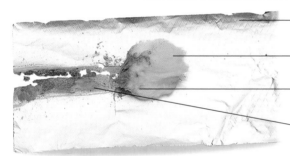

— Unreactive layer of aluminium oxide

— Cotton wool soaked in mercury(II) chloride

— Mercury(II) chloride removes aluminium's oxide layer

— Aluminium reacts with air to reform oxide layer

REMOVING THE OXIDE LAYER
Metallic aluminium, which is used to make kitchen foil and saucepans, seems unreactive. Actually, aluminium is quite high in the activity series. When pure aluminium is exposed to the air, a thin layer of unreactive aluminium oxide rapidly forms on the surfaces, preventing further reaction.

DISPLACEMENT OF COPPER(II) IONS BY ZINC METAL
A **displacement reaction** is one in which **atoms** or ions of one substance take the place of atoms or ions of another. Here, zinc loses electrons to copper ions and displaces copper from a blue solution of copper(II) sulphate. The **products** of this reaction are copper metal and colourless zinc(II) sulphate solution.

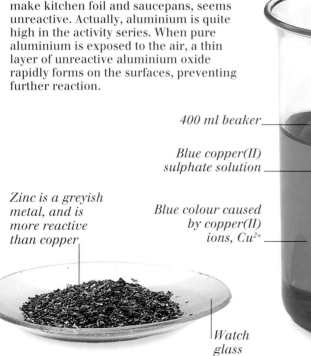

Zinc is a greyish metal, and is more reactive than copper

ZINC METAL

400 ml beaker

Blue copper(II) sulphate solution

Blue colour caused by copper(II) ions, Cu^{2+}

Watch glass

COPPER(II) SULPHATE SOLUTION

Zinc metal dissolves to form zinc(II) ions, Zn^{2+}

Zinc(II) sulphate solution is colourless

Red-brown copper metal forms as it is displaced from solution

ZINC(II) SULPHATE SOLUTION AND METALLIC COPPER

CATHODIC PROTECTION

Sacrificial tubing of more reactive metal | Steel structure | Offshore oil rig

PROTECTION OF OIL RIGS
Many metals corrode when exposed to water and air. To prevent underwater or underground metal pipes from corroding, a more reactive metal may be placed in contact with the pipe. Being more reactive, this metal corrodes in preference to the pipe. This technique, called cathodic protection, is commonly used in oil rigs.

REACTIONS OF METALS WITH DILUTE ACIDS

Acid solutions contain hydrogen ions, H^+, in the form of **hydroxonium ions**, H_3O^+ (see pp. 22-23). Reactive metals in an acid solution donate electrons to hydrogen ions, producing hydrogen gas. Metal atoms become positive ions and dissolve. The more reactive the metal, the faster the reaction proceeds. Some metals are so unreactive that they will only react with hot concentrated acid, and some will not react with acids at all.

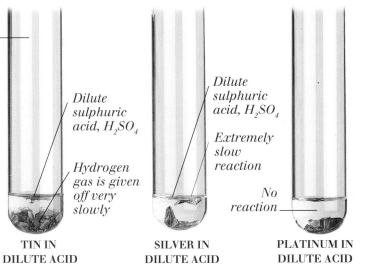

Reaction proceeds fairly rapidly

Bubbles of hydrogen gas, H_2

Zinc, Zn, is a fairly reactive metal

Magnesium, Mg, is a reactive metal

MAGNESIUM IN DILUTE ACID

ZINC IN DILUTE ACID

Test tube

Dilute sulphuric acid, H_2SO_4

Hydrogen gas is given off very slowly

Dilute sulphuric acid, H_2SO_4

Extremely slow reaction

No reaction

TIN IN DILUTE ACID

SILVER IN DILUTE ACID

PLATINUM IN DILUTE ACID

DISPLACEMENT OF SILVER(I) IONS BY COPPER METAL

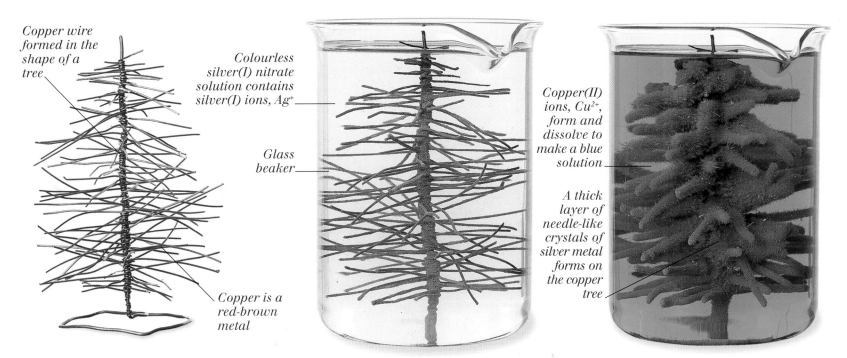

Copper wire formed in the shape of a tree

Colourless silver(I) nitrate solution contains silver(I) ions, Ag^+

Glass beaker

Copper is a red-brown metal

Copper(II) ions, Cu^{2+}, form and dissolve to make a blue solution

A thick layer of needle-like crystals of silver metal forms on the copper tree

COPPER WIRE "TREE"
Here wire made from copper is formed into the shape of a tree. This shape has a large surface area, upon which the reaction can occur.

COPPER TREE IN SILVER NITRATE SOLUTION
When the copper wire is submerged in a solution of silver(I) nitrate, the copper metal loses electrons to the silver(I) ions.

DEPOSITION OF SILVER CRYSTALS
The silver ions are displaced to form silver metal, which coats the copper tree. A blue solution of copper(II) nitrate forms.

Electrochemistry

ELECTRICITY PLAYS A PART in all **chemical reactions**, because all **atoms** consist of **electrically charged** particles (see pp. 10-11). A flow of charged particles is called a current, and is usually carried around a circuit by **electrons**, moved by an electromotive force, or voltage. In **solution**, the charge carriers are **ions**, which are also moved by a voltage. A solution containing ions that conducts current is called an **electrolyte**. There are two basic types of electrochemical system or cell. In an electrolytic cell, two conductors called **electrodes** are dipped in an electrolyte, and connected via an external circuit to a battery or other source of voltage. Such a cell can **decompose** the electrolyte in a process called **electrolysis**. Electrolytic cells are also used in the electroplating of metals. In a voltaic cell, electrodes of two different metals are dipped in an electrolyte. The electrodes produce a voltage that can drive a current between them. Voltaic cells are the basis of common batteries. In both types of cell, the **anode** is the electrode at which **oxidation** occurs, and the **cathode** the one where **reduction** occurs. The cathode is the positive terminal of voltaic cells, but negative in electrolytic cells.

ALKALINE DRY CELL (VOLTAIC)

Electrochemistry is put to use in this **alkaline** dry cell. Powdered zinc metal forms one electrode, while manganese(IV) oxide forms the other. This cell produces electricity at 1.5 volts. Batteries producing 3, 4.5, 6, or 9 volts are made by connecting a series of these cells.

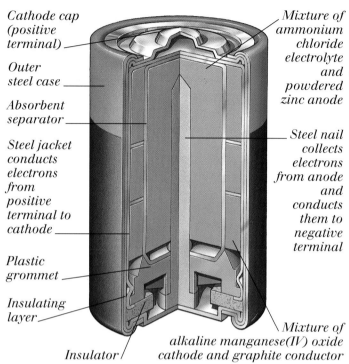

Cathode cap (positive terminal)

Outer steel case

Absorbent separator

Steel jacket conducts electrons from positive terminal to cathode

Plastic grommet

Insulating layer

Insulator

Mixture of ammonium chloride electrolyte and powdered zinc anode

Steel nail collects electrons from anode and conducts them to negative terminal

Mixture of alkaline manganese(IV) oxide cathode and graphite conductor

ELECTROLYSIS

ELECTROLYTIC DECOMPOSITION OF WATER
Passing an electric current through water decomposes it, producing the gases hydrogen and oxygen. A small amount of an **ionic compound** is dissolved in the water to make an electrolyte, into which two electrodes are dipped. The battery removes electrons, e^-, from one electrode, the anode, and pushes them towards the cathode. This is an example of an electrolytic cell.

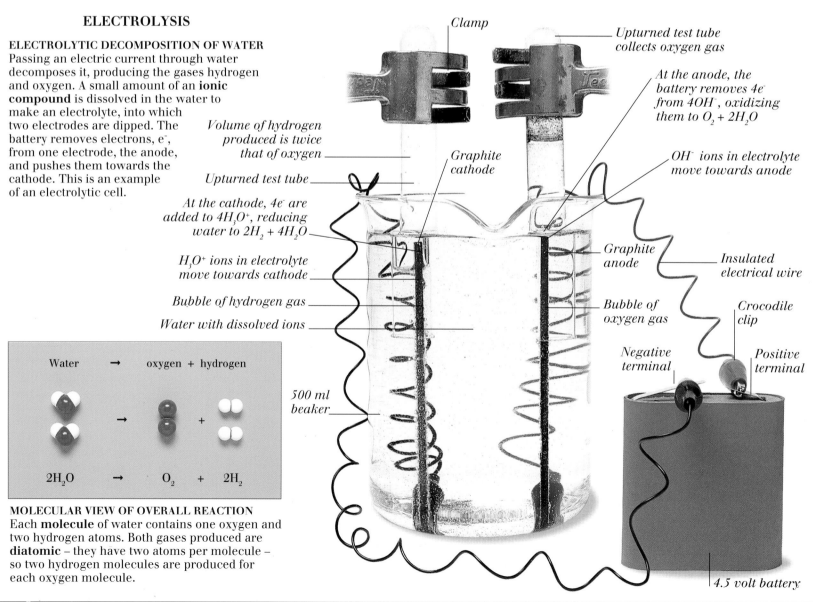

Clamp

Upturned test tube collects oxygen gas

At the anode, the battery removes $4e^-$ from $4OH^-$, oxidizing them to $O_2 + 2H_2O$

OH$^-$ ions in electrolyte move towards anode

Volume of hydrogen produced is twice that of oxygen

Graphite cathode

Upturned test tube

At the cathode, $4e^-$ are added to $4H_3O^+$, reducing water to $2H_2 + 4H_2O$

H_3O^+ ions in electrolyte move towards cathode

Bubble of hydrogen gas

Water with dissolved ions

Graphite anode

Insulated electrical wire

Bubble of oxygen gas

Crocodile clip

Negative terminal

Positive terminal

500 ml beaker

4.5 volt battery

Water	→	oxygen	+	hydrogen

$$2H_2O \rightarrow O_2 + 2H_2$$

MOLECULAR VIEW OF OVERALL REACTION
Each **molecule** of water contains one oxygen and two hydrogen atoms. Both gases produced are **diatomic** – they have two atoms per molecule – so two hydrogen molecules are produced for each oxygen molecule.

PRODUCING A VOLTAGE

When two electrodes of different metals are dipped in an **acidic** solution so that they do not touch each other, an electric voltage is set up between them. This arrangement is called a voltaic cell. If the two electrodes are connected externally by a wire, the voltage causes an

VOLTAIC CELL

electric current to flow. In the voltaic cell below, zinc atoms are oxidized to zinc(II) ions at the anode. Electrons from this oxidation flow through the wire, illuminating the light bulb, to the copper cathode, where hydrogen ions in solution are reduced to hydrogen gas.

Copper cathode (positive terminal of cell)

Crocodile clip

Bubbles of hydrogen gas produced as H+ ions are reduced to H₂ gas

Dilute solution of sulphuric acid, H₂SO₄

Glass dish

1.5 volt light bulb

Bulb holder

Electric circuit

Zinc anode (negative terminal of cell)

Insulated electrical wire

Zinc electrode dissolves in acid

Some bubbles of hydrogen gas here, since zinc undergoes local reaction with acid (see p. 33)

Water molecule

Sulphate ion, SO₄²⁻

Electron

Zinc electrode dissolves

Zinc atoms in electrode

Zinc ion, Zn²⁺, in solution

ZINC ELECTRODE

Zinc atoms in the electrode dissolve in the acid, losing electrons to form **cations**. Oxidation occurs, so this electrode is the anode.

Water molecule

Electron

Copper atoms in electrode

Diatomic hydrogen molecule, H₂

Hydrogen ion, H+, in solution

Sulphate ion, SO₄²⁻

COPPER ELECTRODE

Here, at the cathode, electrons arrive from the zinc anode via the external circuit. They reduce hydrogen ions from the acid, forming hydrogen gas molecules.

Battery removes electrons from copper atoms of anode, forming copper ions

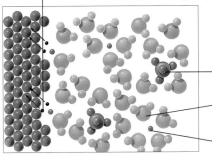

Sulphate ion, SO₄²⁻

Water molecule

Copper ion, Cu²⁺

AT THE COPPER PIPE ANODE

The battery's positive terminal draws electrons from the anode, oxidizing the copper atoms to copper(II) cations. These ions dissolve and move towards the cathode.

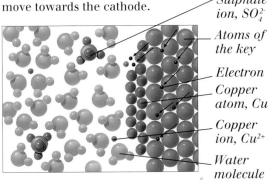

Sulphate ion, SO₄²⁻

Atoms of the key

Electron

Copper atom, Cu

Copper ion, Cu²⁺

Water molecule

AT THE BRASS KEY CATHODE

Copper ions that have moved to the cathode are reduced to copper atoms by electrons from the battery. These atoms build up on the surface of the brass key cathode.

ELECTROPLATING

COPPER PLATING A KEY

In electroplating, a thin layer of one metal is deposited onto the surface of another. The item to be plated is made the cathode in an electrolytic cell. The electrolyte is a solution containing ions of the other metal. Here, a brass key is plated with copper. The copper ions in solution are replenished from a copper anode.

Battery's positive terminal draws electrons from copper anode

Crocodile clip

Copper pipe anode

At the anode, Cu becomes Cu²⁺ + 2e⁻

Negative terminal of battery pushes electrons to brass key cathode

Blue solution of copper(II) sulphate

Copper ions, Cu²⁺, move through solution towards the cathode

4.5 volt battery

Cu²⁺ in solution and 2e⁻ from battery form Cu metal at the cathode

Brass key cathode

600 ml beaker

Rough coating of copper metal

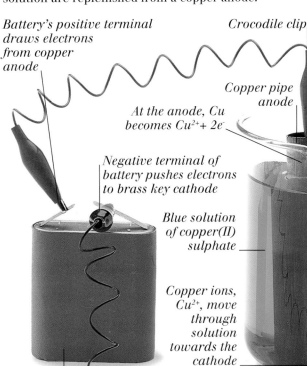

BRASS KEY (BEFORE)

COPPER-PLATED KEY (AFTER)

The alkali metals

THE ELEMENTS OF GROUP 1 of the periodic table (see pp. 12-13) are called the alkali metals. **Atoms** of these **elements** have one outer **electron**. This electron is easily lost, forming singly charged **cations** such as the lithium **ion**, Li⁺. As with all cations, the lithium cation is smaller than the lithium atom. All of the elements in this group are highly **reactive** metals (see pp. 14-15). They react violently with **acids**, and even react with water to form **alkaline solutions** (see pp. 22-23) – hence their group name. The most important element in this group is sodium. Sodium forms many **compounds**, including sodium chloride, or common salt, and sodium hydrogencarbonate, which is used in baking powder. By far the most important compound of sodium in industrial use is sodium hydroxide. It is manufactured in large quantities, mainly by the **electrolysis** of brine (a solution of sodium chloride). Sodium hydroxide is a strong **base**, and it reacts with the fatty acids in fats and oils to produce soap, which is a **salt** (see pp. 24-25).

POSITION IN THE PERIODIC TABLE

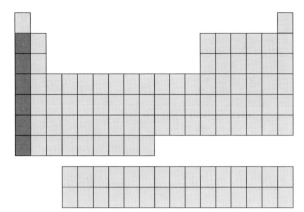

GROUP 1 ELEMENTS
The alkali metals form group 1 of the periodic table. They are (from top): lithium (Li), sodium (Na), potassium (K), rubidium (Rb), caesium (Cs), and francium (Fr).

Potassium is a soft, silvery metal

POTASSIUM METAL

REACTION WITH WATER

Red litmus solution

Glass bowl

The reaction evolves heat

Sodium skims across the surface on a cushion of steam and hydrogen gas

Red litmus begins to turn blue as alkaline sodium hydroxide solution forms

SODIUM IN INDICATOR SOLUTION
A piece of pure sodium metal reacts dangerously with water. Here, red litmus **indicator** is dissolved in the water. Explosive hydrogen gas is given off by the reaction, and the litmus turns blue with the resulting sodium hydroxide solution (above).

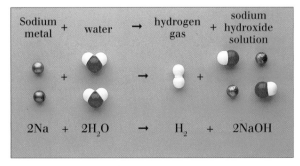

Sodium metal	+	water	→	hydrogen gas	+	sodium hydroxide solution
2Na	+	2H₂O	→	H₂	+	2NaOH

MOLECULAR VIEW
Sodium atoms lose electrons to form sodium cations, Na⁺, which dissolve in water. Water **molecules** each gain an electron and split into a hydroxide **anion**, which dissolves, and a hydrogen atom. Two atoms of hydrogen combine to form hydrogen gas, H₂.

ATOMS AND CATIONS

Atoms of the alkali metals have one electron, which is easily lost, in their outer **electron shell**. The cation formed is much smaller than the atom. Atomic and ionic diameters are given below for the first four alkali metals, measured in picometres (1 picometre, pm, is 10^{-12} m). Electron configurations of the elements are also given.

Atomic diameter 304 pm

Ionic diameter 136 pm

LITHIUM ATOM, 1S² 2S¹

LITHIUM ION, 1S²

Atomic diameter 370 pm

Ionic diameter 194 pm

SODIUM ATOM, 1S² 2S² 2P⁶ 3S¹

SODIUM ION, 1S² 2S² 2P⁶

Ionic diameter 266 pm

Atomic diameter 462 pm

POTASSIUM ATOM, 1S² 2S² 2P⁶ 3S² 3P⁶ 4S¹

POTASSIUM ION, 1S² 2S² 2P⁶ 3S² 3P⁶

Ionic diameter 294 pm

Atomic diameter 492 pm

RUBIDIUM ATOM, 1S² 2S² 2P⁶ 3S² 3P⁶ 3D¹⁰ 4S² 4P⁶ 5S¹

RUBIDIUM ION, 1S² 2S² 2P⁶ 3S² 3P⁶ 3D¹⁰ 4S² 4P⁶

SODIUM HYDROGENCARBONATE

Sodium hydrogencarbonate, NaHCO₃ – also known as bicarbonate of soda – is a weak base that **decomposes** on heating or on reaction with an acid, releasing carbon dioxide gas (see pp. 22-23). This white powder is used as a raising agent in cooking, and is an important ingredient of soda bread.

Sodium hydrogencarbonate decomposes in the heat of the oven, producing carbon dioxide gas

Soda bread

Light texture due to bubbles

Bubbles formed by carbon dioxide

Dough hardens in the oven

SODA BREAD

MANUFACTURE OF SODIUM HYDROXIDE

Much of the sodium hydroxide, NaOH, manufactured is made by the mercury cathode process. This two-stage process begins with the electrolysis of brine, NaCl, to give chlorine gas and pure sodium. The sodium then reacts with water to give sodium hydroxide solution. Mercury is very toxic, and this process is banned in some countries.

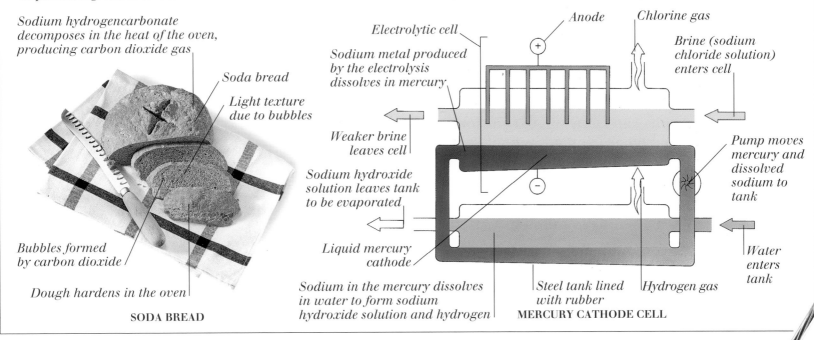

Electrolytic cell

Anode

Chlorine gas

Brine (sodium chloride solution) enters cell

Sodium metal produced by the electrolysis dissolves in mercury

Weaker brine leaves cell

Pump moves mercury and dissolved sodium to tank

Sodium hydroxide solution leaves tank to be evaporated

Liquid mercury cathode

Water enters tank

Sodium in the mercury dissolves in water to form sodium hydroxide solution and hydrogen

Steel tank lined with rubber

Hydrogen gas

MERCURY CATHODE CELL

PRODUCTION OF SOAP

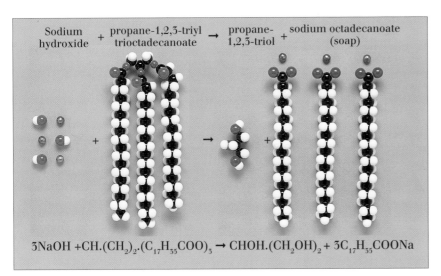

Sodium hydroxide + propane-1,2,3-triyl trioctadecanoate → propane-1,2,3-triol + sodium octadecanoate (soap)

$$3NaOH + CH.(CH_2)_2.(C_{17}H_{35}COO)_3 \rightarrow CHOH.(CH_2OH)_2 + 3C_{17}H_{35}COONa$$

MOLECULAR VIEW

The oil molecule shown consists of three long chain fatty acids linked by propane-1,2,3-triol (glycerol). Sodium hydroxide reacts with the fatty acids from the oil to produce glycerol and the salt sodium octadecanoate.

LABORATORY PREPARATION

When fatty acids – weak acids found in fats and oils – are heated with sodium hydroxide, a strong base, they react to produce a **mixture** of salts. The main product is the salt sodium octadecanoate, C₁₇H₃₅COONa (a soap). Common salt (sodium chloride) helps to separate the soap from the mixture.

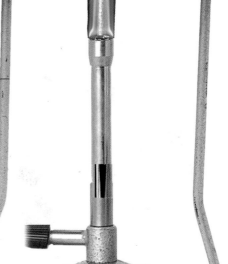

Soap forms as a layer on the top of the mixture

Glass stirrer

250 ml beaker

Oil contains fatty acids

Sodium hydroxide is a corrosive chemical

The mixture is heated

Gauze

Tripod

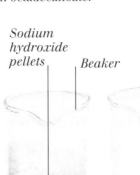
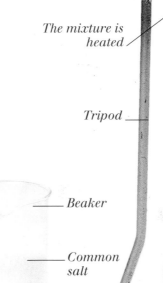

Glass bottle

Sodium hydroxide pellets

Beaker

Olive oil

Beaker

Common salt

The alkaline earth metals

THE ELEMENTS OF THE SECOND GROUP of the periodic table (see pp. 12-13) are called the alkaline earth metals. These **elements** are **reactive**, because their **atoms** easily lose two outer **electrons** to form doubly charged **cations**, such as the calcium **ion**, Ca^{2+}. Hard water, which contains large numbers of dissolved ions, often contains calcium ions. It is formed when slightly **acidic** water flows over rocks containing calcium **salts** such as calcium carbonate. The dissolved calcium salts can come out of **solution** from hard water, forming the scale that blocks kettles and hot water pipes. It is difficult to create a lather with soap when using hard water. In fact, a simple way to measure the hardness of water is to **titrate** it with a soap solution. Calcium **compounds** are an important constituent of mortar, which is used as a cement in bricklaying. Magnesium, another group 2 element, is found in the pigment chlorophyll, which gives green plants their colour. Alkaline earth metals are commonly used in the manufacture of fireworks, and barium is used in hospitals for the production of X-rays of the digestive system.

POSITION IN THE PERIODIC TABLE

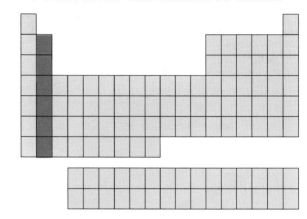

GROUP 2 ELEMENTS
The metals of group 2 of the periodic table are (from top): beryllium (Be), magnesium (Mg), calcium (Ca), strontium (Sr), barium (Ba), and radium (Ra).

MAGNESIUM IN CHLOROPHYLL

Leaf

CHLOROPHYLL IN GREEN PLANTS
Green plants contain large amounts of a vital compound called chlorophyll. It absorbs **energy** from sunlight in a process called **photosynthesis**. The energy is used to make sugars (see pp. 52-53) from carbon dioxide and water.

Green colour is caused by magnesium in chlorophyll pigment

Leaf contains a store of energy built up by photosynthesis

Each cell of the leaf contains chlorophyll

Molecule head

Nitrogen atom

Magnesium atom

MOLECULE OF CHLOROPHYLL
The group 2 element magnesium plays a vital role in the chlorophyll **molecule**. Located at the centre of the porphyrin ring in the head of the molecule, it absorbs light energy as part of the process of photosynthesis.

HARDNESS OF WATER

Burette

COMPARATIVE TITRATION
Hard water contains calcium hydrogencarbonate, $Ca(HCO_3)_2$, or other dissolved salts. These salts increase the amount of soap needed to produce a lather. The hardness of different water samples may be compared by titrating them with a soap solution of fixed **concentration**.

Clamp stand

Soap solution contains a little alcohol to prevent clouding

Clamp

Burette reading is noted

TESTING HARD WATER
A solution of liquid soap in water is slowly added to a sample of hard water. The water is shaken occasionally, and the volume of soap solution is noted when a lather begins to form. Different water samples require different amounts of soap.

Flask is shaken occasionally

Conical flask

Tap

Water sample (tap water)

Water sample (bottled mineral water)

Water sample (rain water)

Porphyrin ring

Phytol side-chain

Tail of molecule

Hydrogen atom

Carbon atom

Covalent bonds

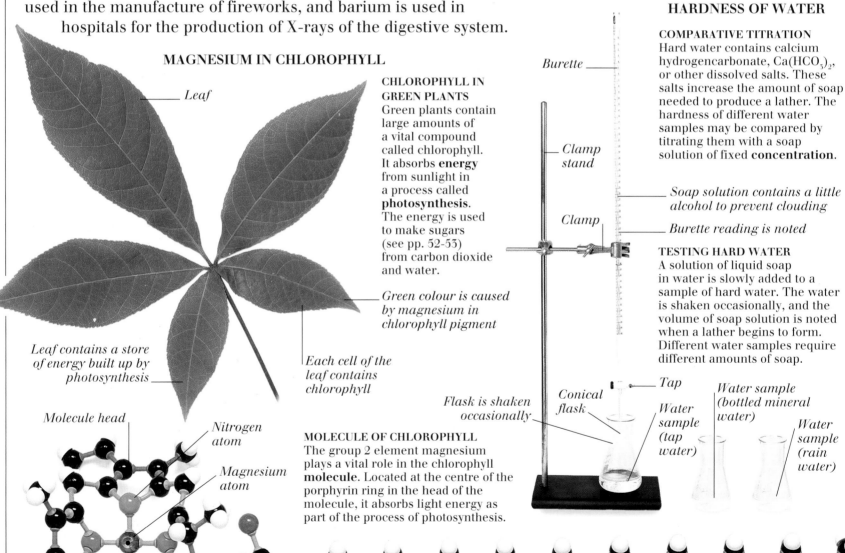

ALKALINE EARTH METALS IN FIREWORKS

Red colour given by strontium salts

Magnesium salts give an intense white colour

CHARACTERISTIC COLOURS
Group 2 elements produce bright colours when heated in a flame (see pp. 54-55). For this reason, compounds of the elements are used in fireworks. As gunpowder in the fireworks burns, electrons in the group 2 atoms absorb heat energy and radiate it out as light of characteristic colours.

BARIUM MEAL

Large intestine   *Skeleton*

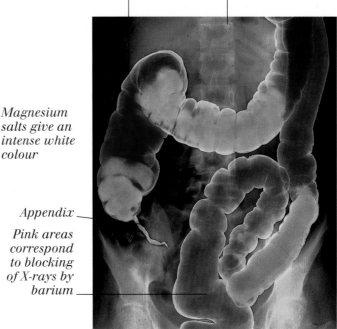

Appendix

Pink areas correspond to blocking of X-rays by barium

X-RAY PHOTOGRAPH OF DIGESTIVE SYSTEM
To obtain an X-ray of the digestive system, a "meal" of barium sulphate, $BaSO_4$, is administered to the patient. X-rays pass through human tissue, but are stopped by atoms of barium.

CALCIUM COMPOUNDS IN MORTAR

PRODUCTION OF MORTAR
Bricklayers' mortar is calcium hydroxide – also known as slaked lime, $Ca(OH)_2$ – dissolved in water, and mixed with sand for bulk. As the mixture dries, the slaked lime **crystallizes** out of solution, and slowly reacts with carbon dioxide in the air to form hard calcium carbonate (see below).

Water evaporates from the mixture

Mortar hardens as it reacts with carbon dioxide in air to form calcium carbonate

250ml beaker

Water

Mould containing wet mortar

Sand and calcium hydroxide, $Ca(OH)_2$, mixed with water

The ingredients are mixed thoroughly

Spatula

Mortar takes the shape of the mould

MOLECULAR VIEW
Calcium ions, Ca^{2+}, and hydroxide ions, OH^-, form when slaked lime dissolves in water. Carbon dioxide, CO_2, combines with the ions as water leaves the mixture. This reaction is also the basis of a test for carbon dioxide (see pp. 54-55).

Calcium hydroxide	+	carbon dioxide	→	calcium carbonate	+	water

$Ca(OH)_2$ + CO_2 → $CaCO_3$ + H_2O

Transition metals

THE TRANSITION METALS MAKE UP MOST of the periodic table (see pp. 12-13). Some of the **elements** are very familiar – for example, gold and silver are used in jewellery, copper is used in electrical wiring and water pipes, and tungsten forms the filaments of incandescent light bulbs. **Transition metals** share many properties – for example, they all have more than one **oxidation number**. In **compounds**, chromium commonly has oxidation numbers of +2, +3, or +6. Like most transition metals, it forms coloured **ions** in **solution**, such as the chromate(VI) and the dichromate(VI) ions. Copper also exhibits typical transition metal behaviour – it forms brightly coloured compounds and **complex ions**. Perhaps the most important of the transition metals is iron. It is the most widely used of all metals, and is usually **alloyed** with precise amounts of carbon and other elements to form steel. Around 760 million tonnes of steel are produced per year world-wide, most of it by the basic oxygen process. Chromium is used in stainless steel alloys, and as a shiny protective plating on other metals.

POSITION IN THE PERIODIC TABLE

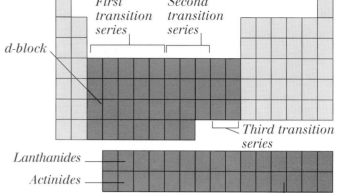

First transition series

Second transition series

d-block

Third transition series

Lanthanides

Actinides

f-block

D- AND F-BLOCK ELEMENTS
Most of the transition metals lie in the d-block of the periodic table. The lanthanides and actinides, in the f-block, are also transition metals.

THREE D-BLOCK TRANSITION METALS

Gold is very unreactive

Silver compounds are used in photographic film

Platinum is often used as a catalyst

GOLD **SILVER** **PLATINUM**

COPPER – A TRANSITION METAL

Copper(II) hydroxide can be prepared by adding a strong alkali to copper(II) salts

COPPER(II) NITRATE

Copper(II) nitrate is hygroscopic, which means that it absorbs water from the air

THREE TRANSITION METAL COMPOUNDS

Like many chromium compounds, chromium(III) oxide, Cr_2O_3, is used as a pigment

CHROMIUM(III) OXIDE

COMPOUNDS OF COPPER
Like most of the transition metals, copper forms brightly coloured compounds. All of the compounds shown here are of copper(II), and they all contain the ion Cu^{2+}.

COPPER(II) HYDROXIDE

Copper(II) oxide, CuO, is used as a catalyst in a number of reactions

COPPER(II) OXIDE

Chromium(VI) oxide, CrO_3, is highly poisonous

CHROMIUM(VI) OXIDE

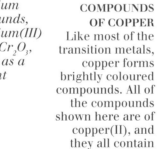

Copper(II) carbonate contains the carbonate radical, CO_3^{2-}

COPPER(II) CARBONATE

This form of lead(II) oxide is called litharge

LEAD(II) OXIDE

This blue-green sample of copper(II) chloride contains water of crystallization

Red-brown copper turnings

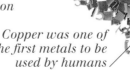

Copper was one of the first metals to be used by humans

COPPER(II) CHLORIDE

COPPER METAL

MANUFACTURE OF STEEL

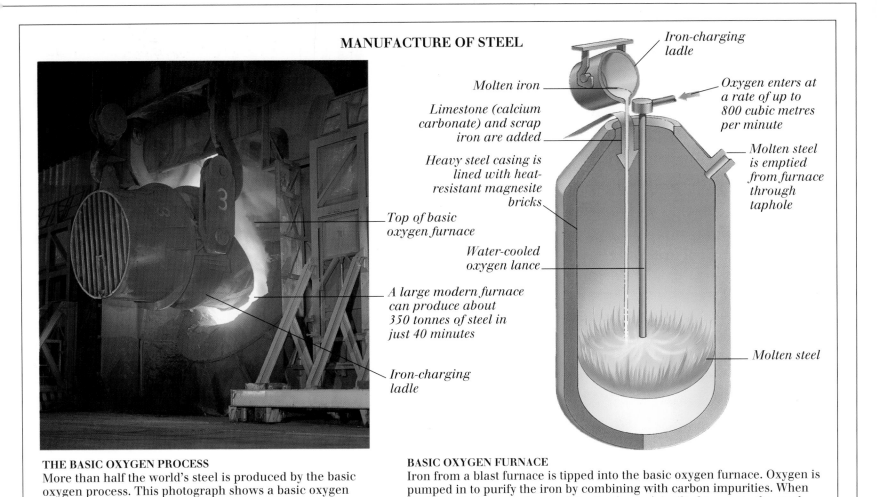

Iron-charging ladle

Molten iron

Limestone (calcium carbonate) and scrap iron are added

Heavy steel casing is lined with heat-resistant magnesite bricks

Top of basic oxygen furnace

Water-cooled oxygen lance

A large modern furnace can produce about 350 tonnes of steel in just 40 minutes

Iron-charging ladle

Oxygen enters at a rate of up to 800 cubic metres per minute

Molten steel is emptied from furnace through taphole

Molten steel

THE BASIC OXYGEN PROCESS
More than half the world's steel is produced by the basic oxygen process. This photograph shows a basic oxygen furnace (right) being charged, or filled, with molten iron.

BASIC OXYGEN FURNACE
Iron from a blast furnace is tipped into the basic oxygen furnace. Oxygen is pumped in to purify the iron by combining with carbon impurities. When the "blow" of oxygen is complete, the furnace is tilted to empty the steel.

CHROMATE IONS IN A REVERSIBLE REACTION

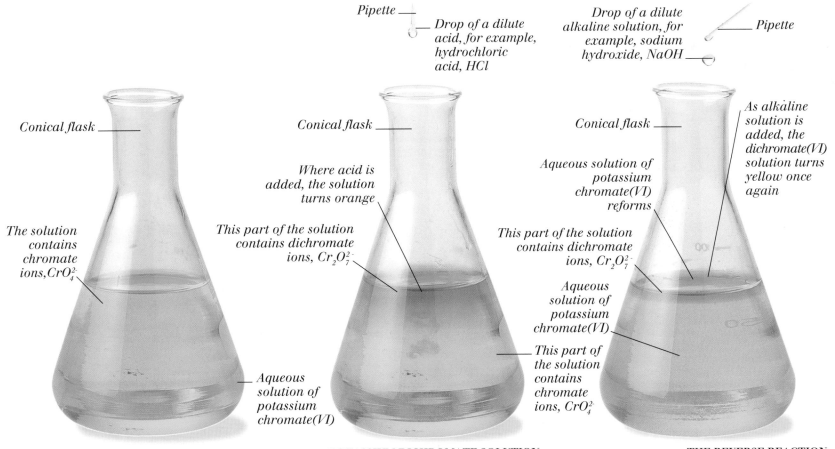

Pipette

Drop of a dilute acid, for example, hydrochloric acid, HCl

Drop of a dilute alkaline solution, for example, sodium hydroxide, NaOH

Pipette

Conical flask

The solution contains chromate ions, CrO_4^{2-}

Conical flask

Where acid is added, the solution turns orange

This part of the solution contains dichromate ions, $Cr_2O_7^{2-}$

Aqueous solution of potassium chromate(VI)

Conical flask

Aqueous solution of potassium chromate(VI) reforms

This part of the solution contains dichromate ions, $Cr_2O_7^{2-}$

Aqueous solution of potassium chromate(VI)

This part of the solution contains chromate ions, CrO_4^{2-}

As alkaline solution is added, the dichromate(VI) solution turns yellow once again

POTASSIUM CHROMATE SOLUTION
When dissolved in water, the compound potassium chromate(VI), K_2CrO_4, has a bright yellow colour. Chromium in the compound has an oxidation number of +6.

POTASSIUM DICHROMATE SOLUTION
Adding an **acid** to the solution moves the position of the **equilibrium**. Two chromate(VI) ions combine to produce the dichromate(VI) ion, $Cr_2O_7^{2-}$, and water.

THE REVERSE REACTION
The addition of more water or an **alkaline** solution will push the **reversible reaction** in the direction of the original **reactants**. A yellow solution of chromate(VI) ions forms once more.

Carbon, silicon, and tin

GROUP 14 OF THE periodic table (see pp. 14-15) contains the
elements carbon, silicon, and tin. Carbon is a non-metal that is
the basis of **organic** chemistry (see pp. 50-53). It occurs in three
distinct forms, or **allotropes**. In the most recently discovered of
these, called the fullerenes, carbon **atoms** join together in a
hollow spherical cage. The other, more familiar, allotropes of
carbon are graphite and diamond. All of the elements in group
14 form sp **hybrid orbitals** (see p. 17). In particular, sp^3 hybrid
orbitals give a tetrahedral structure to many of the **compounds**
of these elements. Silicon is a semi-metal that is used in
electronic components. It is found naturally in many types
of rock, including quartz, which consists of silicon(IV) oxide.
Quartz is the main constituent of sand, which is used to make
glass. Tin is a metallic element. It is not very useful in its pure
form, because it is soft and weak. However, combined with
other metals, it forms useful **alloys**, such as solder and bronze.

POSITION IN THE PERIODIC TABLE

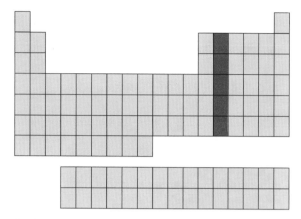

GROUP 14 ELEMENTS
Group 14 of the periodic table consists of (top to
bottom): carbon(C), silicon (Si), germanium (Ge),
tin (Sn), and lead (Pb).

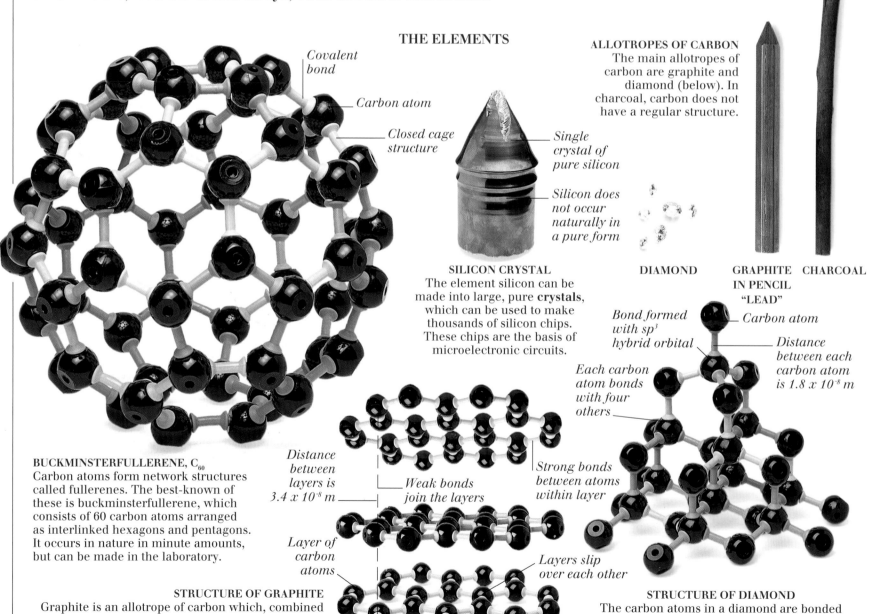

THE ELEMENTS

Covalent bond

Carbon atom

Closed cage structure

ALLOTROPES OF CARBON
The main allotropes of
carbon are graphite and
diamond (below). In
charcoal, carbon does not
have a regular structure.

Single crystal of pure silicon

Silicon does not occur naturally in a pure form

SILICON CRYSTAL
The element silicon can be
made into large, pure **crystals**,
which can be used to make
thousands of silicon chips.
These chips are the basis of
microelectronic circuits.

DIAMOND

GRAPHITE IN PENCIL "LEAD"

CHARCOAL

Bond formed with sp^3 hybrid orbital

Carbon atom

Distance between each carbon atom is 1.8 x 10^{-8} m

Each carbon atom bonds with four others

BUCKMINSTERFULLERENE, C$_{60}$
Carbon atoms form network structures
called fullerenes. The best-known of
these is buckminsterfullerene, which
consists of 60 carbon atoms arranged
as interlinked hexagons and pentagons.
It occurs in nature in minute amounts,
but can be made in the laboratory.

Distance between layers is 3.4 x 10^{-8} m

Weak bonds join the layers

Strong bonds between atoms within layer

Layer of carbon atoms

Layers slip over each other

STRUCTURE OF GRAPHITE
Graphite is an allotrope of carbon which, combined
with various clays, forms the "lead" of pencils. The
carbon atoms in graphite form layers that are loosely
bound together, and slip easily over each other.

STRUCTURE OF DIAMOND
The carbon atoms in a diamond are bonded
in a very strong structure. Each carbon
atom is bound directly to four others, which
sit at the corners of a tetrahedron.

SP³ HYBRIDIZATION

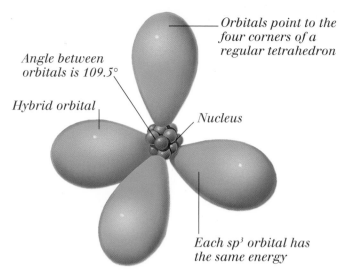

Orbitals point to the four corners of a regular tetrahedron

Angle between orbitals is 109.5°

Hybrid orbital

Nucleus

Each sp³ orbital has the same energy

FORMATION OF SP³ ORBITALS

The elements in group 14 of the periodic table have one s- and three p-orbitals in their outer **electron shells**. These combine to form four sp³ hybrid orbitals in many of the compounds of the elements.

QUARTZ AND GLASS

Quartz crystal

Pure quartz is clear

QUARTZ

Quartz is the most abundant rock type on Earth. It consists mainly of the compound silicon(IV) oxide.

GLASS

Glass is made from molten sand, which consists mainly of quartz (above). Sodium and calcium **salts** are added to lower the melting point of the sand. The glass can be coloured by adding impurities such as barium carbonate and iron(III) oxide.

BROWN GLASS BOTTLE **GREEN GLASS BOTTLE**

Sodium carbonate lowers the melting point of sand, but makes the glass soluble in water

Calcium carbonate lowers the melting point of sand without making the glass soluble in water

Sand consists of grains of quartz

Barium carbonate gives glass a brown colour

Iron(III) oxide gives glass a green colour

SODIUM CARBONATE   **CALCIUM CARBONATE**   **SAND (MAINLY QUARTZ)**   **BARIUM CARBONATE**   **IRON(III) OXIDE**

ALLOYS OF TIN

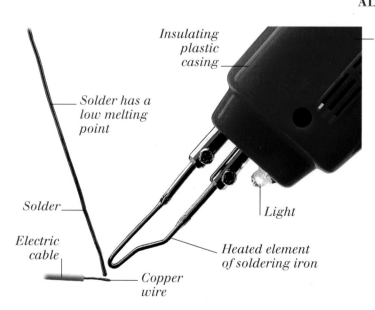

Insulating plastic casing

Electric soldering iron

Solder has a low melting point

Solder

Electric cable

Copper wire

Light

Heated element of soldering iron

Bronze is an alloy of copper and tin

Green surface layer (patina) forms as copper oxidizes

Ancient bronze statue of a horse

SOLDER

The most convenient way to connect wires and components permanently in electric circuits is to use solder. Solder is a soft alloy of tin and lead that has a low melting point (200-300°C).

BRONZE

First made about 5000 years ago, bronze is an alloy of tin and copper (see pp. 40-41). It is easily cast when molten, but very hard-wearing when solidified.

Nitrogen and phosphorus

NITROGEN AND PHOSPHORUS are the two most important **elements** in group 15 of the periodic table (see pp. 12-13). Phosphorus, which is solid at room temperature, occurs in two forms, or **allotropes**, called white and red phosphorus. Nitrogen is a gas at room temperature, and makes up about 78% of air (see pp. 8-9). Fairly pure nitrogen can be prepared in the laboratory by removing oxygen, water vapour, and carbon dioxide from air. By far the most important **compound** of nitrogen is ammonia (see pp. 30-31), of which over 80 million tonnes are produced each year worldwide. Used in the manufacture of fertilizers, explosives, and nitric acid, ammonia is produced industrially by the Haber process, for which nitrogen and hydrogen are the raw materials. Ammonia forms a positive **ion** called the ammonium ion (NH_4^+) that occurs in **salts**, where it acts like a metal **cation** (see pp. 14-15). Ammonia can be prepared in the laboratory by heating an ammonium salt with an **alkali**, such as calcium hydroxide.

THE POSITION OF NITROGEN AND PHOSPHORUS IN THE PERIODIC TABLE

Nitrogen (N, uppermost) and phosphorus (P) are non-metals that belong to group 15 of the periodic table.

Glass bowl

Sticks of white phosphorus

Water

White phosphorus is a white, waxy solid

White phosphorus melts at 44.1°C

WHITE PHOSPHORUS

ALLOTROPES OF PHOSPHORUS

There are two common allotropes of phosphorus. White phosphorus reacts violently with air, so it is kept in water, in which it does not dissolve. It changes slowly to the noncrystalline red form, which is chemically less reactive.

Red phosphorus melts at about 600°C

RED PHOSPHORUS

Watch glass

Red phosphorus is a red powder at room temperature

PREPARATION OF NITROGEN FROM AIR

Nitrogen is the most abundant gas in the air. Other gases that make up more than 1% of the air are oxygen (about 20%) and water vapour (0-4%). Air is passed through sodium hydroxide **solution**, which dissolves the small amounts of carbon dioxide present. It is then passed through concentrated sulphuric acid to remove water vapour, and, finally, over heated copper metal to remove oxygen. The result is almost pure nitrogen.

Air is pumped slowly through the apparatus from this glass tube

Rubber bung

Round-bottomed flask

Air is dried by the sulphuric acid

Solution of sodium hydroxide

Rubber bung

Concentrated sulphuric acid

Glass tube

Turnings of copper metal

Gas flame heats the copper turnings

Bunsen burner

Air hole open to give hot blue flame

Rubber bung

Clamp

Hot copper turnings combine with oxygen in the air to form copper(II) oxide

Gas sample will still contain small amounts of noble gases, such as argon

Delivery tube

Nitrogen is an invisible gas at room temperature

Gas displaces water from boiling tube

Almost pure nitrogen

Bubbles of gas

Water

Delivery tube is bent

Upturned boiling tube collects gas

THE HABER PROCESS

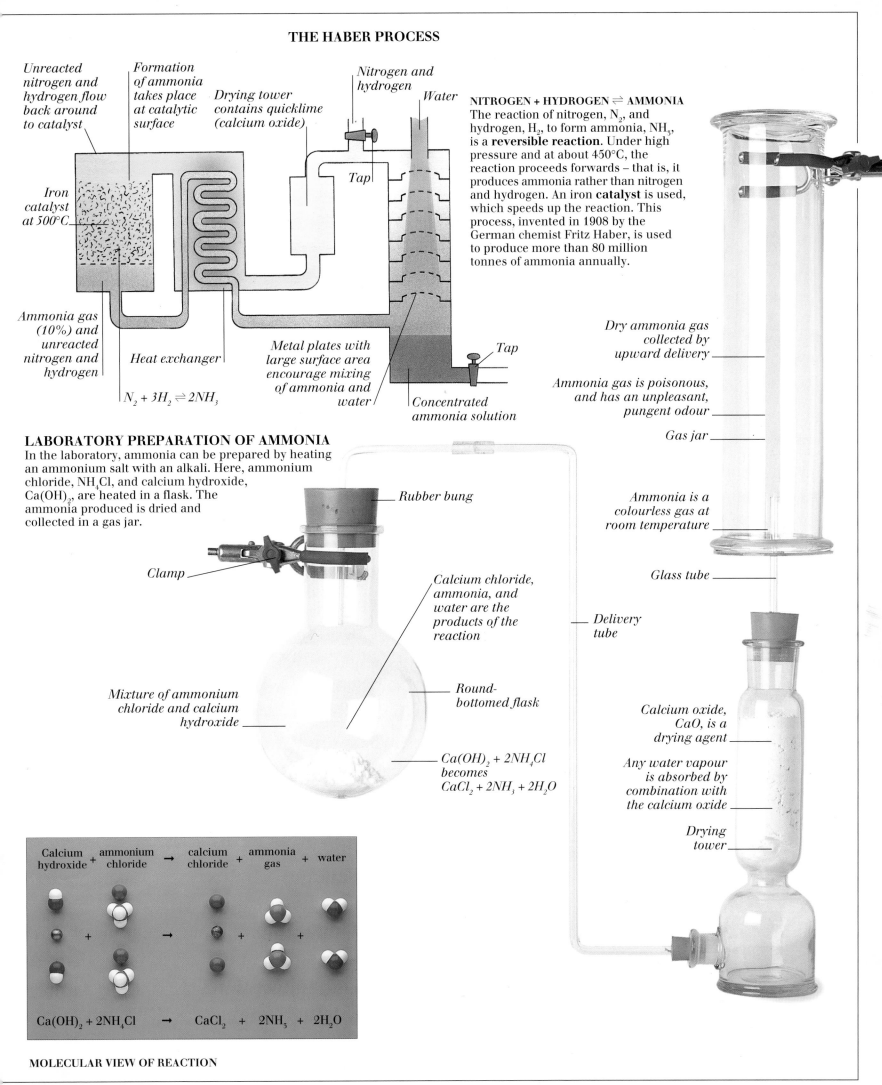

Unreacted nitrogen and hydrogen flow back around to catalyst

Formation of ammonia takes place at catalytic surface

Drying tower contains quicklime (calcium oxide)

Nitrogen and hydrogen

Water

Iron catalyst at 500°C

Tap

Ammonia gas (10%) and unreacted nitrogen and hydrogen

Heat exchanger

$N_2 + 3H_2 \rightleftharpoons 2NH_3$

Metal plates with large surface area encourage mixing of ammonia and water

Tap

Concentrated ammonia solution

NITROGEN + HYDROGEN ⇌ AMMONIA
The reaction of nitrogen, N_2, and hydrogen, H_2, to form ammonia, NH_3, is a **reversible reaction**. Under high pressure and at about 450°C, the reaction proceeds forwards – that is, it produces ammonia rather than nitrogen and hydrogen. An iron **catalyst** is used, which speeds up the reaction. This process, invented in 1908 by the German chemist Fritz Haber, is used to produce more than 80 million tonnes of ammonia annually.

Dry ammonia gas collected by upward delivery

Ammonia gas is poisonous, and has an unpleasant, pungent odour

Gas jar

Ammonia is a colourless gas at room temperature

Glass tube

LABORATORY PREPARATION OF AMMONIA
In the laboratory, ammonia can be prepared by heating an ammonium salt with an alkali. Here, ammonium chloride, NH_4Cl, and calcium hydroxide, $Ca(OH)_2$, are heated in a flask. The ammonia produced is dried and collected in a gas jar.

Rubber bung

Clamp

Calcium chloride, ammonia, and water are the products of the reaction

Delivery tube

Mixture of ammonium chloride and calcium hydroxide

Round-bottomed flask

$Ca(OH)_2 + 2NH_4Cl$ becomes $CaCl_2 + 2NH_3 + 2H_2O$

Calcium oxide, CaO, is a drying agent

Any water vapour is absorbed by combination with the calcium oxide

Drying tower

Calcium hydroxide + ammonium chloride → calcium chloride + ammonia gas + water

$Ca(OH)_2 + 2NH_4Cl \rightarrow CaCl_2 + 2NH_3 + 2H_2O$

MOLECULAR VIEW OF REACTION

Oxygen and sulphur

THE TWO MOST IMPORTANT **elements** in group 16 of the periodic table are oxygen and sulphur. Oxygen, a gas at **STP**, is vital to life, and is one of the most abundant elements on Earth. It makes up 21% by volume of dry air (see pp. 8-9). In the laboratory, oxygen is easily prepared by the **decomposition** of hydrogen peroxide. Oxygen is involved in **burning** – it relights a glowing wooden splint, and this is one test for the gas. Sulphur occurs in several different structural forms, known as **allotropes**. The most stable allotrope at room temperature is rhombic sulphur, in which sulphur exists in the form of rings, each containing eight atoms. One important **compound** of sulphur is hydrogen sulphide. It has a pungent smell like that of rotten eggs, and can be prepared by reacting dilute **acids** with metal sulphides. Sodium thiosulphate is another important sulphur compound, used as a fixer in the development of photographic images.

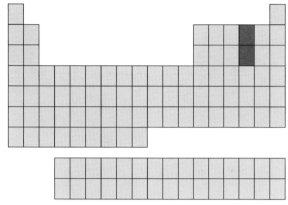

GROUP 16 ELEMENTS
Oxygen (top) and sulphur are in group 16 of the periodic table. They are both non-metallic elements, which form a wide range of compounds.

TEST FOR OXYGEN

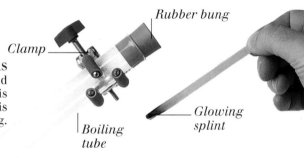

Clamp — *Rubber bung*

Boiling tube

Glowing splint

OXYGEN GAS
A tube full of oxygen gas, produced in the reaction on the left, is sealed. A previously lit splint is extinguished, but left glowing.

— *Oxygen gas*

RELIT SPLINT
The glowing splint (above) relights in the oxygen. Burning, or combustion, is defined as a rapid combination of a substance with oxygen. It is a **redox reaction** (see pp. 20-21).

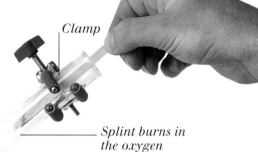

Clamp

Boiling tube

Splint burns in the oxygen

PREPARATION OF OXYGEN

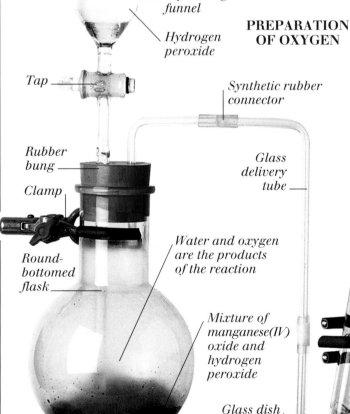

Separating funnel

Hydrogen peroxide

Tap

Synthetic rubber connector

Rubber bung

Clamp

Glass delivery tube

Round-bottomed flask

Water and oxygen are the products of the reaction

Mixture of manganese(IV) oxide and hydrogen peroxide

Glass dish

Oxygen gas fills the boiling tube

Clamp

Upturned boiling tube

Water

CATALYTIC DECOMPOSITION OF HYDROGEN PEROXIDE
The decomposition of hydrogen peroxide to oxygen and water normally occurs very slowly. The addition of a **catalyst** of manganese(IV) oxide to hydrogen peroxide speeds up the reaction.

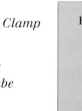

Hydrogen peroxide	→	water	+	oxygen

$$2H_2O_2 \rightarrow 2H_2O + O_2$$

MOLECULAR VIEW
Hydrogen peroxide exists as **molecules**, each consisting of two hydrogen and two oxygen **atoms**. Every two molecules of hydrogen peroxide produce one molecule of oxygen. Water is the other **product**.

ALLOTROPES OF SULPHUR

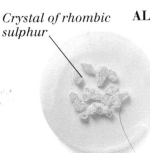

Crystal of rhombic sulphur

Plastic sulphur forms crystals on cooling

Crystals formed by rapidly cooling molten sulphur

Watch glass

Atom of sulphur

Covalent bond

POWDERED SULPHUR
In the laboratory, sulphur is usually supplied as a powder. Each grain of the powder is a crystal of rhombic sulphur (right).

RHOMBIC (α) SULPHUR
The most stable allotrope of sulphur at room temperature is rhombic sulphur, also known as alpha (α) sulphur.

PLASTIC SULPHUR
If molten sulphur is cooled by plunging it into cold water, yellow or brown plastic sulphur, which is non-crystalline, forms.

MONOCLINIC (β) SULPHUR
Needle-like crystals of monoclinic (β, or beta) sulphur slowly revert to the rhombic form at temperatures below 95.5°C (203.9°F).

SULPHUR RINGS
Monoclinic and rhombic sulphur both contain crown-shaped molecules of eight sulphur atoms.

LABORATORY PREPARATION OF HYDROGEN SULPHIDE

The pungent gas hydrogen sulphide, H_2S, is normally prepared by the action of a dilute acid on metal sulphides. In this case, the reactants are hydrochloric acid, HCl, and iron(II) sulphide, FeS.

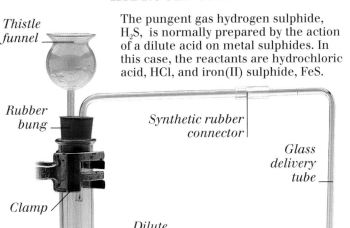

Thistle funnel

Rubber bung

Synthetic rubber connector

Glass delivery tube

Clamp

Dilute hydrochloric acid

Gas jar

Mixture of hydrochloric acid and iron(II) sulphide

Round-bottomed flask

Hydrogen sulphide gas

PREPARATION OF SODIUM THIOSULPHATE

The compound sodium thiosulphate, $Na_2S_2O_3$, is normally combined with water of crystallization. It is used as photographers' fixer, "hypo". Sodium thiosulphate is prepared by heating a **suspension** of sulphur in a sodium sulphite, Na_2SO_3, **solution**.

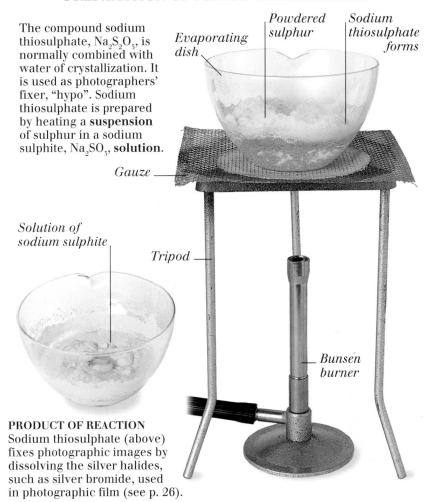

Evaporating dish

Powdered sulphur

Sodium thiosulphate forms

Gauze

Solution of sodium sulphite

Tripod

Bunsen burner

PRODUCT OF REACTION
Sodium thiosulphate (above) fixes photographic images by dissolving the silver halides, such as silver bromide, used in photographic film (see p. 26).

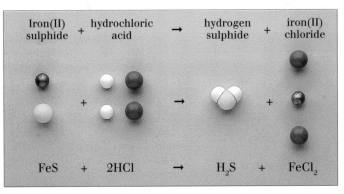

| Iron(II) sulphide | + | hydrochloric acid | → | hydrogen sulphide | + | iron(II) chloride |

| FeS | + | 2HCl | → | H_2S | + | $FeCl_2$ |

MOLECULAR VIEW
Hydrogen ions in hydrochloric acid combine with the sulphur from iron(II) sulphide. Hydrogen sulphide molecules have a similar shape to water molecules (see p. 6).

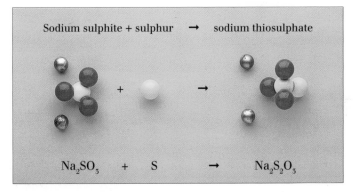

| Sodium sulphite | + | sulphur | → | sodium thiosulphate |

| Na_2SO_3 | + | S | → | $Na_2S_2O_3$ |

MOLECULAR VIEW
Sulphur in the sulphite ion has an **oxidation number** of +4. In the thiosulphate ion produced by the reaction above, sulphur(IV) has been **oxidized** to sulphur(VI), while elemental sulphur has been **reduced** to an oxidation state of -2.

The halogens

THE ELEMENTS OF GROUP 17 of the periodic table (see pp. 12-13) are called the halogens. **Atoms** of these **elements** are just one **electron** short of a full outer **electron shell**. Halogen atoms easily gain single electrons, forming singly charged halide **anions** such as the fluoride **ion**, F⁻. This makes the elements in this group highly **reactive** – some halogens will even react with the **noble gases** under extreme conditions. Chlorine, the most important halogen, is a greenish-yellow **diatomic** gas at room temperature. Chlorine can be prepared in the laboratory by the **oxidation** of hydrochloric acid. Small amounts of chlorine are added to water in swimming pools, and to some water supplies, to kill bacteria. Simple tests may be used to measure the amount of dissolved chlorine. If the **concentration** of chlorine is too high, it can endanger human health – if it is too low, it might not be effective. One important chlorine **compound** is sodium chlorate(I), the main ingredient of domestic bleach. Other halogen compounds include **CFCs** (chlorofluorocarbons). CFCs deplete, or break down, the ozone layer in the upper atmosphere, allowing harmful radiation from the Sun to reach the Earth's surface.

POSITION IN THE PERIODIC TABLE

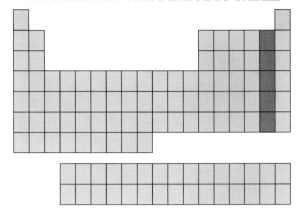

GROUP 17 ELEMENTS
The halogens form group 17 of the periodic table. They are (top to bottom): fluorine (F), chlorine (Cl), bromine (Br), iodine (I), and astatine (At).

PREPARATION OF CHLORINE
Chlorine gas, Cl_2, is prepared in the laboratory by the oxidation of hydrochloric acid, HCl, using manganese(IV) oxide, MnO_2. The chlorine produced contains some water vapour, but is dried by passing it through concentrated sulphuric acid, H_2SO_4. In order to prevent this acid being sucked back into the reaction vessel, an empty dreschel bottle is placed between the acid bottle and the vessel to act as an anti-suck-back device (see p. 23). The dry gas is collected in a gas jar. Chlorine gas is poisonous.

BROMINE AND IODINE
The element bromine is a red liquid at room temperature, though it vaporizes easily, producing a brown vapour. Iodine is a violet solid at room temperature, which sublimes (turns to vapour without passing through a liquid phase) when warmed.

Gas jar

Gas jar

Violet iodine vapour produced by warming solid iodine

Bromine vapour is brownish

Blue-black crystals of solid iodine

Liquid bromine

BROMINE

IODINE

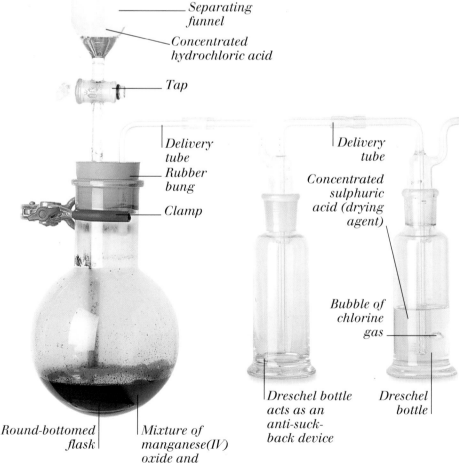

Separating funnel

Concentrated hydrochloric acid

Tap

Delivery tube

Rubber bung

Clamp

Round-bottomed flask

Mixture of manganese(IV) oxide and hydrochloric acid

Delivery tube

Concentrated sulphuric acid (drying agent)

Bubble of chlorine gas

Dreschel bottle acts as an anti-suck-back device

Dreschel bottle

CHLORINE IN WATER

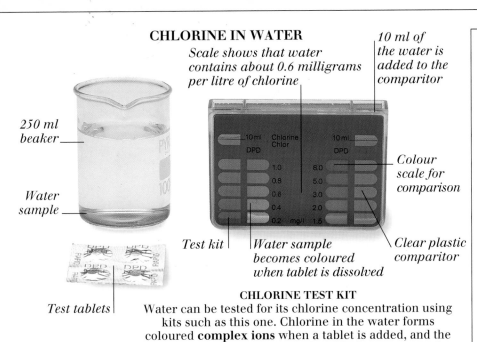

250 ml beaker

Water sample

Scale shows that water contains about 0.6 milligrams per litre of chlorine

10 ml of the water is added to the comparitor

Colour scale for comparison

Water sample becomes coloured when tablet is dissolved

Clear plastic comparitor

Test kit

Test tablets

CHLORINE TEST KIT
Water can be tested for its chlorine concentration using kits such as this one. Chlorine in the water forms coloured **complex ions** when a tablet is added, and the intensity of the colour reveals the chlorine concentration.

BLEACHING

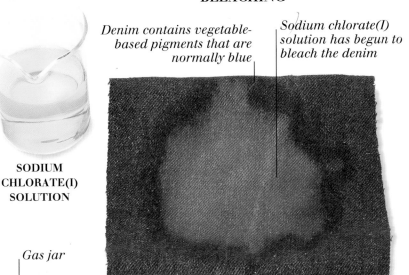

Denim contains vegetable-based pigments that are normally blue

Sodium chlorate(I) solution has begun to bleach the denim

SODIUM CHLORATE(I) SOLUTION

BLEACHING ACTION OF SODIUM CHLORATE(I)
Sodium chlorate(I), NaOCl, is an industrially important chlorine compound that is a strong oxidizing agent. It bleaches pigments by giving up its oxygen to them, making them colourless.

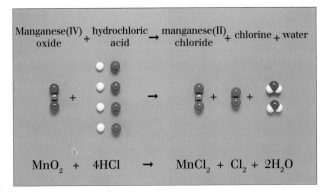

Gas jar

Chlorine is a greenish-yellow gas at room temperature

$$Manganese(IV)\ oxide + hydrochloric\ acid \rightarrow manganese(II)\ chloride + chlorine + water$$

$$MnO_2 + 4HCl \rightarrow MnCl_2 + Cl_2 + 2H_2O$$

MOLECULAR VIEW
Four units of the hydrochloric acid are oxidized by each **molecule** of manganese(IV) oxide. The manganese(IV) is **reduced** to manganese(II).

OZONE DEPLETION REACTIONS
CFCs, synthetic **organic** compounds containing chlorine and fluorine atoms, have been used in packaging and some aerosol cans. Released into the atmosphere, CFCs lose chlorine atoms. These atoms **catalyse** reactions that damage the ozone layer, which shields the Earth from harmful solar radiation.

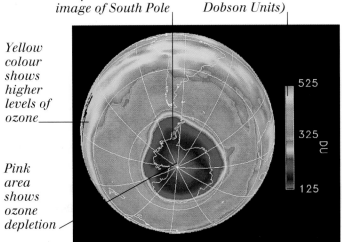

Computer-enhanced image of South Pole

Level of ozone (scale in Dobson Units)

Yellow colour shows higher levels of ozone

Pink area shows ozone depletion

525

325

DU

125

DEPLETION OF THE OZONE LAYER

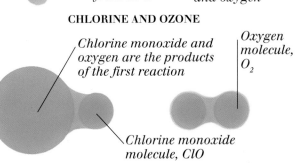

Ozone molecule, O_3

Chlorine atom released from CFC

Ozone reacts with chlorine to form chlorine monoxide and oxygen

CHLORINE AND OZONE

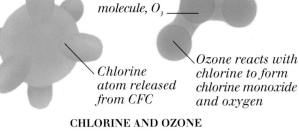

Chlorine monoxide and oxygen are the products of the first reaction

Oxygen molecule, O_2

Chlorine monoxide molecule, ClO

CHLORINE MONOXIDE AND OXYGEN

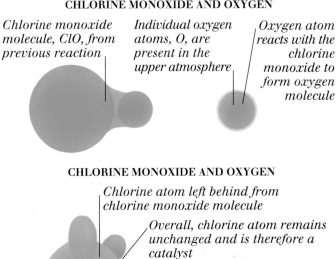

Chlorine monoxide molecule, ClO, from previous reaction

Individual oxygen atoms, O, are present in the upper atmosphere

Oxygen atom reacts with the chlorine monoxide to form oxygen molecule

CHLORINE MONOXIDE AND OXYGEN

Chlorine atom left behind from chlorine monoxide molecule

Overall, chlorine atom remains unchanged and is therefore a catalyst

Oxygen molecule, O_2

CHLORINE AND OXYGEN

Organic chemistry 1

ORGANIC CHEMISTRY IS THE study of carbon **compounds**, although it normally excludes carbon dioxide and **salts** such as calcium carbonate (see pp. 24-25). There are more carbon-based compounds than compounds based on all the other **elements** put together. This is because carbon **atoms** easily bond to each other, forming long chains and rings that include single bonds, double bonds (see p. 17), and triple bonds. **Hydrocarbons** are **molecules** containing only carbon and hydrogen. There are three main families of hydrocarbons based on carbon chains, called **alkanes**, **alkenes**, and **alkynes** (right). Ethyne is the simplest alkyne, with two carbon atoms. Most carbon compounds occur in different structural forms, or **isomers**. For example, the hydrocarbon butene has two isomers that differ in the position of the double bond. Crude oil is a **mixture** (see pp. 8-9) of long-chain hydrocarbons, which is separated industrially in a fractionating tower, and cracked (heated with a **catalyst**) to produce more useful short-chain compounds.

Particles of soot

Ethyne burns in a flame, producing water vapour and carbon dioxide

Ethyne is a colourless gas

Glass tube

Calcium carbide, CaC₂, is a brown ionic solid

Watch glass

FAMILIES OF HYDROCARBONS
Alkanes have only single bonds in the chain of carbon atoms. Alkenes have at least one double bond in the chain, while alkynes have a triple bond.

ALKANES

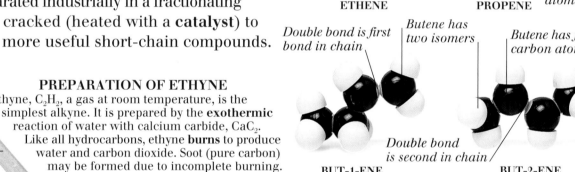

Tetrahedral shape *Carbon atom* *Hydrogen atom*

Single bond

METHANE **ETHANE**

Propane has three carbon atoms *Butane has four carbon atoms*

PROPANE **BUTANE**

ALKENES

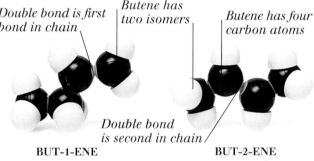

Ethene has two carbon atoms *Double bond*

Propene has three carbon atoms

ETHENE **PROPENE**

Double bond is first bond in chain *Butene has two isomers* *Butene has four carbon atoms*

Double bond is second in chain

BUT-1-ENE **BUT-2-ENE**

ALKYNES

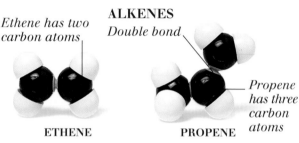

Triple bond *Propyne exists in one form only*

ETHYNE

Triple bond *Single bond*

Triple bond is second bond in carbon chain

PROPYNE

Butyne has two isomers

BUT-2-YNE

Triple bond is first bond in carbon chain

PREPARATION OF ETHYNE
Ethyne, C₂H₂, a gas at room temperature, is the simplest alkyne. It is prepared by the **exothermic** reaction of water with calcium carbide, CaC₂. Like all hydrocarbons, ethyne **burns** to produce water and carbon dioxide. Soot (pure carbon) may be formed due to incomplete burning.

Rubber bung *Clamp stand*

Boiling tube

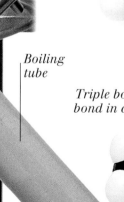

Clamp

CaC₂ and H₂O become C₂H₂ and Ca(OH)₂

BUT-1-YNE

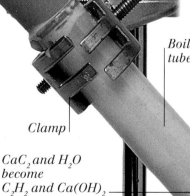

CALCIUM CARBIDE
Calcium carbide, CaC₂, is an **ionic** solid that contains the Ca^{2+} and C_2^{2-} ions. In ethyne, the **product** of the reaction, carbon and hydrogen atoms are **covalently bound**.

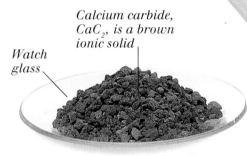

calcium carbide	+	water	→	ethyne	+	calcium hydroxide

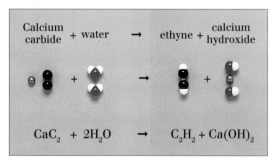

$$CaC_2 + 2H_2O \rightarrow C_2H_2 + Ca(OH)_2$$

MOLECULAR VIEW
Carbon atoms from calcium carbide combine with hydrogen atoms from water molecules to form ethyne.

Clamp

Cotton wool

Products may contain pure carbon, deposited on cotton wool

Rubber bung

Crude oil

Catalytic cracking takes place at the surface of the pot pieces

Delivery tube

Porous pot pieces

Crude oil is a mixture of hydrocarbons

Bunsen flame heats the oil

Products of cracking are hydrocarbons with shorter chains

Products may include hydrogen gas

Upturned test tube collects gases

Clamp

CATALYTIC CRACKING
In this laboratory set-up, a mixture of long-chain hydrocarbons is vaporized and passed over pieces of porous pot. The long hydrocarbons attach to the pieces and **decompose** into smaller molecules. The pot acts as a catalyst.

500 ml beaker

Water

FRACTIONAL DISTILLATION

FRACTIONAL DISTILLATION
Crude oil is made up of a mixture of hydrocarbons. This mixture is separated into fractions (groups of hydrocarbons with similar boiling points) by a process called fractional **distillation**. This process takes place in a fractionating tower. The oil is vaporized, and each fraction condenses to a liquid at a different **temperature**.

Naphtha (a mixture of hydrocarbons used for many applications) emerges here

Kerosine (paraffin oil) used as aircraft oil and for domestic heating

Condensed gases (reflux) run down inside of tower

Furnace

Crude oil vaporizes in furnace

Pump

Some of the residue goes to be cracked (above)

Refinery gas escapes at the top of the tower

Condenser

Refinery gas contains methane, ethane, propane, and butane

Water from condenser

Gasoline (light hydrocarbons used for petrol)

Fractionating tower is typically 40 m tall

110°C

150°C

190°C

280°C

300°C

Fractions to be processed for transport fuel

Steam is pumped in to heat unvaporized oil

Residue contains long-chain hydrocarbons, including bitumen for roads and wax for candles

Long-chain alkane → nonane + propane + ethyne + carbon (soot) + hydrogen gas

$$C_{15}H_{52} \rightarrow C_9H_{20} + C_5H_8 + C_2H_2 + C + H_2$$

MOLECULAR VIEW
In the catalytic cracking of oil, hydrocarbon chains, shown here as 15 carbon atoms long, break into smaller chains with between 2 and 9 carbons. This is a molecular model of a general reaction. In reality, many other similar reactions are also likely to occur.

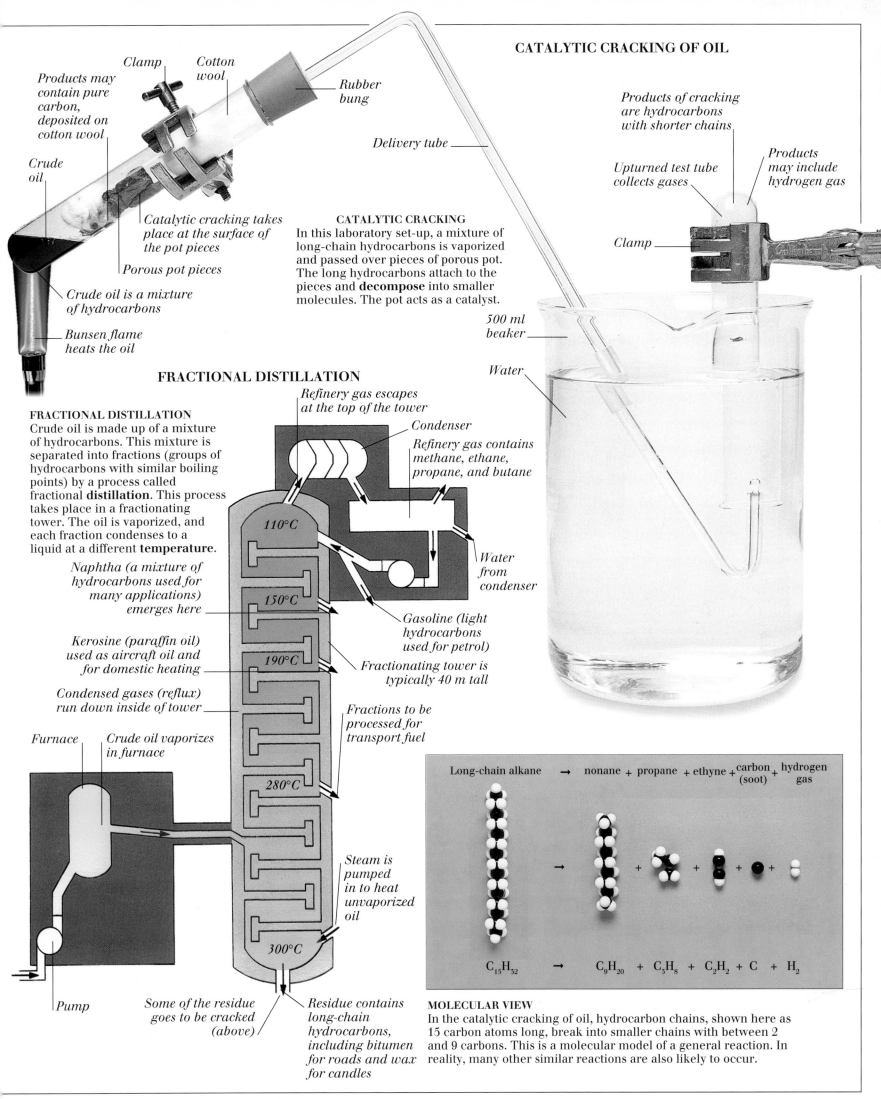

Organic chemistry 2

THE CHEMISTRY OF CARBON is called **organic** chemistry. Simple organic **molecules** (see pp. 50-51) are based on chains of carbon **atoms**. Carbon atoms are very versatile at bonding, and can form very large and complicated molecules. Small organic molecules often join together to form larger ones. For example, glucose, a simple sugar or monosaccharide, is a small organic molecule. Two saccharide units join to form a disaccharide, such as sucrose. Large numbers of sugar units can join to form polysaccharides such as starch (see p. 27). The process of joining large numbers of identical molecules together is called **polymerization**. The polymers that result are commonplace both in synthetic products and in nature. Plastics, such as nylon and PVC, are polymers, and much more complicated polymers form the basis of life. Hæmoglobin is a large organic molecule responsible for carrying oxygen in red blood cells. DNA is a giant molecule that holds the genetic code in all living organisms. This code is created from patterns of four small molecules called bases, which are arranged along the famous double helix structure.

SUCROSE CRYSTALS

Sugars are **carbohydrates**. Sucrose (see p. 27) is the chemical name for ordinary household sugar. In this beaker, **crystals** of sucrose have formed from an **aqueous solution** of household sugar.

String suspended in beaker

Aqueous solution of sugar

Sucrose crystals

Crystals grow from solution around string

Glass beaker

Glass rod

Nylon drawn out as a long thread

Solution of 1,6-diaminohexane in water

250 ml beaker

Nylon forms where two solutions meet

Layers do not mix because hexane does not dissolve in water

Solution of hexanedioic acid in hexane

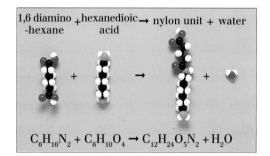

FORMATION OF NYLON

LABORATORY PREPARATION
Nylon is a polymer that is formed from two organic monomers. The form of nylon shown here is made by the synthesis (joining) of the monomers hexanedioic acid and 1,6-diaminohexane.

1,6 diamino-hexane + hexanedioic acid → nylon unit + water

$$C_6H_{16}N_2 + C_6H_{10}O_4 \rightarrow C_{12}H_{24}O_5N_2 + H_2O$$

MOLECULAR VIEW
A unit of nylon is made from one molecule of each monomer (above). Each unit reacts again with one monomer at each end, eventually forming the polymer nylon. A nylon molecule may comprise hundreds of such units.

PLASTICS

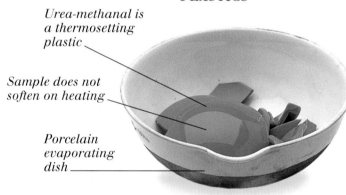

Urea-methanal is a thermosetting plastic

Sample does not soften on heating

Porcelain evaporating dish

THERMOSETTING PLASTICS
Thermosetting plastics are moulded when first made, and harden upon cooling. They cannot be softened again by heating.

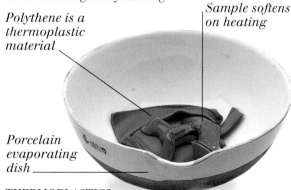

Polythene is a thermoplastic material

Sample softens on heating

Porcelain evaporating dish

THERMOPLASTICS
Some plastics soften on heating. They can be remoulded while hot, then allowed to cool and harden. Polythene is an example of such a thermoplastic.

POLYMERIZATION

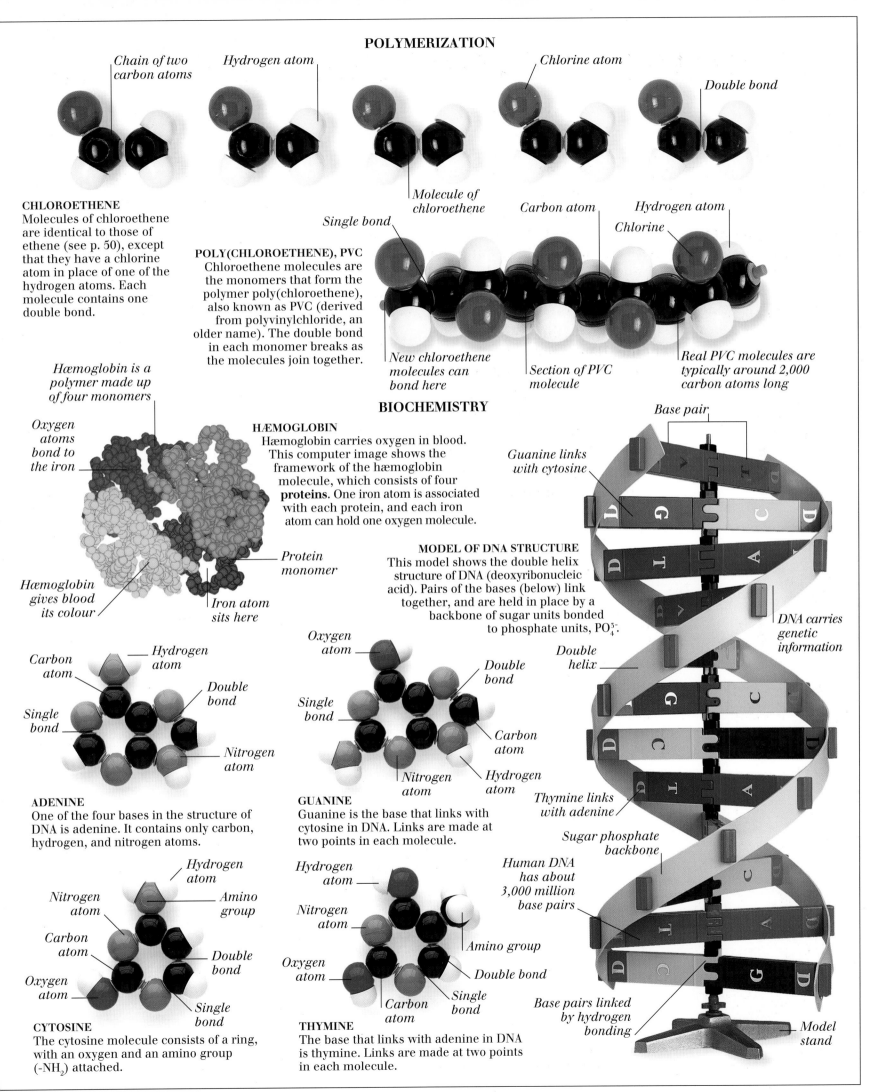

Chain of two carbon atoms

Hydrogen atom

Chlorine atom

Double bond

Molecule of chloroethene

CHLOROETHENE
Molecules of chloroethene are identical to those of ethene (see p. 50), except that they have a chlorine atom in place of one of the hydrogen atoms. Each molecule contains one double bond.

Single bond

Carbon atom

Hydrogen atom

Chlorine

POLY(CHLOROETHENE), PVC
Chloroethene molecules are the monomers that form the polymer poly(chloroethene), also known as PVC (derived from polyvinylchloride, an older name). The double bond in each monomer breaks as the molecules join together.

New chloroethene molecules can bond here

Section of PVC molecule

Real PVC molecules are typically around 2,000 carbon atoms long

BIOCHEMISTRY

Hæmoglobin is a polymer made up of four monomers

Oxygen atoms bond to the iron

HÆMOGLOBIN
Hæmoglobin carries oxygen in blood. This computer image shows the framework of the hæmoglobin molecule, which consists of four **proteins**. One iron atom is associated with each protein, and each iron atom can hold one oxygen molecule.

Protein monomer

Hæmoglobin gives blood its colour

Iron atom sits here

Base pair

Guanine links with cytosine

MODEL OF DNA STRUCTURE
This model shows the double helix structure of DNA (deoxyribonucleic acid). Pairs of the bases (below) link together, and are held in place by a backbone of sugar units bonded to phosphate units, PO_4^{3-}.

DNA carries genetic information

Double helix

Carbon atom

Hydrogen atom

Double bond

Single bond

Nitrogen atom

ADENINE
One of the four bases in the structure of DNA is adenine. It contains only carbon, hydrogen, and nitrogen atoms.

Oxygen atom

Double bond

Single bond

Carbon atom

Nitrogen atom

Hydrogen atom

GUANINE
Guanine is the base that links with cytosine in DNA. Links are made at two points in each molecule.

Thymine links with adenine

Sugar phosphate backbone

Human DNA has about 3,000 million base pairs

Hydrogen atom

Nitrogen atom

Amino group

Carbon atom

Double bond

Oxygen atom

Single bond

CYTOSINE
The cytosine molecule consists of a ring, with an oxygen and an amino group (-NH$_2$) attached.

Hydrogen atom

Nitrogen atom

Oxygen atom

Amino group

Double bond

Carbon atom

Single bond

THYMINE
The base that links with adenine in DNA is thymine. Links are made at two points in each molecule.

Base pairs linked by hydrogen bonding

Model stand

Chemical analysis

THERE ARE MANY SITUATIONS, from geological surveys to forensic investigations, that call for the chemical analysis of unknown substances. The substances being analysed may be present only in tiny amounts, and may be **mixtures** of many different **compounds**. Separation techniques such as **chromatography** (see pp. 8-9) are often the starting point in an analysis. Simple laboratory tests may follow – these normally identify one part of a compound at a time. For example, flame tests are used to identify **cations** of metallic **elements** in a compound, and **radicals** may be identified by heating the compound to **decompose** it, thereby releasing signifying gases. Many simple laboratory tests are performed on **aqueous solutions** of the unknown substance. The substance is crushed and dissolved in water, and other **solutions**, such as ammonium hydroxide or silver nitrate, are added. The colour of any **precipitate** formed indicates the presence of a specific **ion**. In contrast, mass **spectrometry** is a highly complex but very powerful testing technique. The sample to be tested is vaporized, then ionized. The ions are separated by a strong magnetic field and identified according to their **electric charge** and mass.

FLAME TEST

A sample of an unknown compound is held on the end of a platinum wire in a Bunsen burner flame. Specific colours in the flame indicate the presence of certain metals.

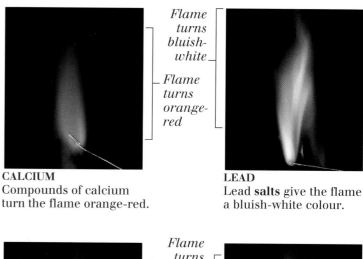

Flame turns bluish-white

Flame turns orange-red

CALCIUM
Compounds of calcium turn the flame orange-red.

LEAD
Lead **salts** give the flame a bluish-white colour.

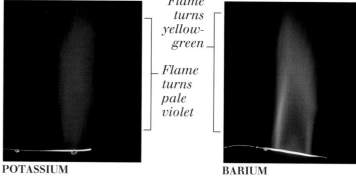

Flame turns yellow-green

Flame turns pale violet

POTASSIUM
Compounds of potassium turn the flame pale violet.

BARIUM
Barium salts turn the flame yellow-green.

TEST FOR A CARBONATE OR HYDROGENCARBONATE

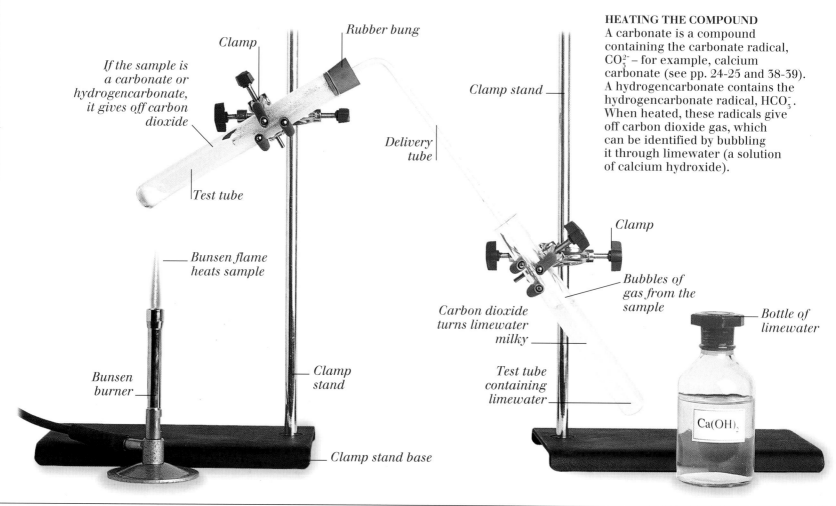

If the sample is a carbonate or hydrogencarbonate, it gives off carbon dioxide

Clamp

Rubber bung

Test tube

Bunsen flame heats sample

Bunsen burner

Clamp stand

Clamp stand base

Clamp stand

Delivery tube

Carbon dioxide turns limewater milky

Test tube containing limewater

Clamp

Bubbles of gas from the sample

Bottle of limewater

Ca(OH)₂

HEATING THE COMPOUND

A carbonate is a compound containing the carbonate radical, CO_3^{2-} – for example, calcium carbonate (see pp. 24-25 and 38-39). A hydrogencarbonate contains the hydrogencarbonate radical, HCO_3^-. When heated, these radicals give off carbon dioxide gas, which can be identified by bubbling it through limewater (a solution of calcium hydroxide).

TESTING FOR CATIONS

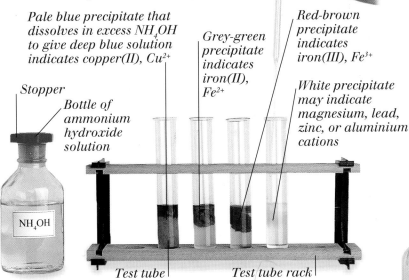

Test tube

Sample of unknown compound dissolved in water

Test tube rack

ACTION OF AMMONIUM HYDROXIDE

Many different tests are used to identify **cations** in an unknown compound. One simple test is carried out on a pure solution of the unknown compound in water. A dilute solution of ammonium hydroxide, NH_4OH, is added to the test solution. If a gelatinous precipitate forms, any cations present can often be identified by the colour of the precipitate.

Pale blue precipitate that dissolves in excess NH_4OH to give deep blue solution indicates copper(II), Cu^{2+}

Grey-green precipitate indicates iron(II), Fe^{2+}

Red-brown precipitate indicates iron(III), Fe^{3+}

White precipitate may indicate magnesium, lead, zinc, or aluminium cations

Stopper

Bottle of ammonium hydroxide solution

NH_4OH

Test tube

Test tube rack

TEST RESULTS

A few solutions have been tested, and precipitates have formed in the test tubes. From the colour of the precipitates, the metal cations present in the samples have been identified.

MASS SPECTROMETER

In a mass spectrometer, the sample to be tested is vaporized, then converted into ions and shot into a curved tube. A magnetic field in the tube deflects those ions with a specific mass and charge into a detector. Changing the magnetic field strength allows a mass spectrum – an analysis of all the ions present – to be built up, from which the the test substance can be accurately identified.

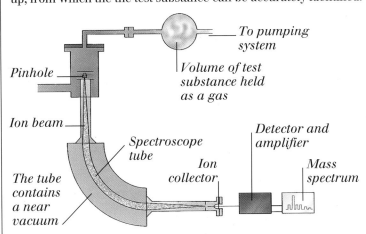

To pumping system

Volume of test substance held as a gas

Pinhole

Ion beam

Spectroscope tube

Ion collector

Detector and amplifier

Mass spectrum

The tube contains a near vacuum

TESTING FOR ANIONS

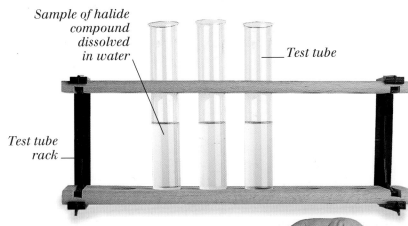

Sample of halide compound dissolved in water

Test tube

Test tube rack

ACTION OF SILVER NITRATE SOLUTION

Of the many different tests used to identify **anions** in an unknown compound, the addition of aqueous silver nitrate, $AgNO_3$, to an aqueous solution of the compound is often the first step. If halide ions – ions of the halogens (see pp. 48-49) – are present in solution, a coloured precipitate forms.

Dropper

White precipitate indicates that the compound contains chloride ions

Pale yellow precipitate indicates that the compound contains bromide ions

Yellow precipitate indicates that the compound contains iodide ions

Bottle of silver nitrate solution

Stopper

$AgNO_3$

Test tube rack

Test tube

TEST RESULTS

Here, solutions of compounds containing ions of the halogens chlorine, iodine, and bromine have been tested. Salts containing ions of the halogens are called halides. Precipitates have formed in the test tubes.

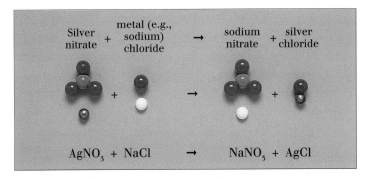

| Silver nitrate | + | metal (e.g., sodium) chloride | → | sodium nitrate | + | silver chloride |

$$AgNO_3 + NaCl \rightarrow NaNO_3 + AgCl$$

MOLECULAR VIEW OF REACTION

A **double decomposition reaction** takes place between silver nitrate and the halide salt in solution, and insoluble silver halides form. Silver halides are used in photography (see p. 26).

Useful data

IMPORTANT IONS AND RADICALS

NAME	FORMULA AND CHARGE
Hydrogen	H^+
Sodium	Na^+
Potassium	K^+
Magnesium	Mg^{2+}
Calcium	Ca^{2+}
Aluminium	Al^{3+}
Iron(II)	Fe^{2+}
Iron(III)	Fe^{3+}
Copper(I)	Cu^+
Copper(II)	Cu^{2+}
Silver(I)	Ag^+
Zinc	Zn^{2+}
Ammonium	NH_4^+
Hydroxonium	H_3O^+
Oxide	O^{2-}
Sulphide	S^{2-}
Fluoride	F^-
Chloride	Cl^-
Bromide	Br^-
Iodide	I^-
Hydroxide	OH^-
Carbonate	CO_3^{2-}
Hydrogencarbonate	HCO_3^-
Nitrate(V)	NO_3^-
Sulphate(VI)	SO_4^{2-}

COMMON NAMES AND FORMULAE OF IMPORTANT COMPOUNDS

COMMON NAME	CHEMICAL NAME	FORMULA
Water	Hydrogen oxide	H_2O
Salt	Sodium chloride	$NaCl$
Bicarbonate of soda	Sodium hydrogencarbonate	$NaHCO_3$
Washing soda	Sodium carbonate decahydrate	$Na_2CO_3.10H_2O$
Household bleach	Sodium chlorate(I)	$NaOCl$
Methylated spirits	Methanol	CH_3OH
Alcohol	Ethanol	C_2H_5OH
Vinegar	Ethanoic acid	$CH_3.COOH$
Vitamin C	Ascorbic acid	$C_4H_5O_4.CHOH.CH_2OH$
Aspirin	Acetylsalicylic acid	$C_6H_4.COOCH_3.COOH$
White sugar	Sucrose	$C_6H_{11}O_5.O.C_6H_{11}O_5$
Limestone/chalk	Calcium carbonate	$CaCO_3$
Plaster of Paris	Calcium sulphate hemihydrate	$CaSO_4.\frac{1}{2}H_2O$
Rust	Hydrated iron(III) oxide	$Fe_2O_3.xH_2O^*$

Exact number of water molecules varies.

NAMES AND STRUCTURES OF COMMON PLASTICS

COMMON NAME OF PLASTIC	PROPER NAME	REPEATED UNIT (MONOMER)
Polythene	Poly(ethene)	Ethene, C_2H_4
PVC or polyvinylchloride	Poly(chloroethene)	Chloroethene, C_2H_3Cl
Polystyrene	Poly(phenylethene)	Phenylethene, $C_2H_3.C_6H_5$
Acrylic	Poly(propenonitrile)	Propenonitrile, $C_2H_2.CH_3.CN$
PTFE or Teflon®	Poly(tetrafluoroethene)	Tetrafluoroethene, C_2F_4

DISCOVERY OF ELEMENTS

ELEMENT NAME	DISCOVERED*	ORIGIN OF NAME
Carbon, C	known since ancient times	Latin *carbo*, charcoal
Gold, Au	known since ancient times	Old English *geolo*, yellow; Latin *aurum*, gold
Sulphur, S	known since ancient times	Latin *sulfur*, brimstone
Platinum, Pt	16th century	Spanish *platina*, little silver
Cobalt, Co	1735 by Georg Brandt	German *kobold*, goblin
Hydrogen, H	1766 by Henry Cavendish	Greek *hydro-* and *genes*, water-maker
Chlorine, Cl	1774 by Carl Wilhelm Scheele	Greek *chloros*, greenish-yellow
Tungsten, W	1783 by Juan José and Fausto Elhuyar	Swedish *tung*, heavy, and *sten*, stone; German *wolfram*
Chromium, Cr	1797 by Louis-Nicolas Vauquelin	Greek *chroma*, colour
Bromine, Br	1826 by Antoine-Jérôme Balard	Greek *bromos*, stench
Helium, He	1868 by Pierre Janssen and Joseph Norman Lockyer	Greek *helios*, the Sun
Unnilquadium, Unq	1964 (in USSR) and 1969 (in USA)	Latin for 104, the element's atomic number**

Generally refers to when the pure substance was first isolated – its recognition as an element often came later.
**Because of disputes over the discovery of elements with atomic numbers 104–109, their names are yet to be finalized.*

MELTING AND BOILING POINTS OF SOME ELEMENTS

ELEMENT	MELTING POINT		BOILING POINT	
	°C	°F	°C	°F
Mercury	-39	-38	357	675
Helium	-272	-458	-269	-452
Tungsten	3,410	6,170	5,555	10,031
Nitrogen	-210	-346	-196	-321
Sodium	98	208	883	1,621
Oxygen	-219	-362	-183	-297
Bromine	-7	19	59	138
Iron	1,535	2,795	2,862	5,184
Carbon	3,550	6,420	4,827	8,720
Gold	1,063	1,945	2,970	5,379

THE MOST ABUNDANT ELEMENTS IN THE EARTH'S CRUST

ELEMENT	MASS (%)
Oxygen	49.13
Silicon	26.00
Aluminium	7.45
Iron	4.20
Calcium	3.25
Sodium	2.40
Potassium	2.35
Magnesium	2.35
Hydrogen	1.00
Others	1.87

Glossary

ACID: A **compound** containing hydrogen that can donate **protons** (hydrogen **ions**, H⁺). In **aqueous solution**, the protons associate with water **molecules** to form **hydroxonium ions**, H_3O^+.

ACTIVATION ENERGY: The least **energy** required for a particular **chemical reaction** to take place. Typically, it is supplied as heat energy – striking a match produces heat to start the match **burning**.

ALKALI: A **base** that is soluble in water. When it dissolves, it produces hydroxide **ions**, OH^-.

ALKANE: A **hydrocarbon** that has only single bonds between its carbon **atoms** – for example, ethane.

ALKENE: A **hydrocarbon** such as ethene that has one or more double bonds between its carbon **atoms**.

ALKYNE: A **hydrocarbon** such as propyne that has one or more triple bonds between its carbon **atoms**.

ALLOTROPES: Forms of the same **element** with different **molecular** or **crystalline** structures. Diamond and graphite are allotropes of carbon.

ALLOY: A **mixture** of a metal with other metals or non-metals in certain proportions, prepared when they are molten. Bronze is an alloy of the metals copper and tin, while steel is an alloy of iron and the non-metal carbon.

ANHYDROUS: Describing a substance that has lost its **water of crystallization**. Adding water rehydrates an anhydrous substance.

ANION: An **ion** or **radical** with a negative **electric charge** – for example, the fluoride ion, F^-, and the sulphate radical, SO_4^{2-}. Anions are attracted to the **anode** during **electrolysis**.

ANODE: In an **electrochemical cell**, the **electrode** where **oxidation**

RHOMBIC (α) SULPHUR CRYSTALS

occurs. The anode is the positive terminal in an electrolytic cell, but negative in a voltaic cell.

AQUEOUS SOLUTION: A **solution** in which the **solvent** is water.

ATOM: The smallest part of an **element** that retains its chemical identity. Overall electrically neutral, atoms consist of negatively charged **electrons** that surround a central, positively charged **nucleus**.

ATOMIC FORCE MICROSCOPE: A device used to produce images of **atoms**, which even very powerful light microscopes cannot resolve. A probe scans a solid surface, closely following its contours. A computer converts the probe's motion into an image of the surface atoms.

ATOMIC NUMBER: The number of **protons** in the **nucleus** of an **atom**, which is unique to each **element**.

BASE: A **compound** that can accept **protons** (hydrogen **ions**, H⁺) to neutralize **acids**, producing a **salt** and water.

AQUEOUS SOLUTION OF AMMONIA

BURNING: Also called combustion, burning is the rapid combination of a substance with oxygen. It is an **exothermic** reaction.

CARBOHYDRATE: An **organic compound**, such as a sugar, that contains the **elements** carbon, hydrogen, and oxygen only.

CATALYST: A substance that increases the **rate** of a **chemical reaction**, but is itself unchanged at the end of the reaction.

CATHODE: In an **electrochemical cell**, the **electrode** where **reduction** occurs. The cathode is the negative terminal in an electrolytic cell, but the positive one in a voltaic cell.

CATION: An **ion** or **radical** with a positive **electric charge**. Metals readily form cations, such as the copper(II) ion, Cu^{2+}. Cations are attracted to the **cathode** during **electrolysis**.

CFC: Abbreviation for chlorofluorocarbon. Any **compound** formed by replacing some or all of the hydrogen **atoms** of a **hydrocarbon** with chlorine and

fluorine atoms. CFCs released in the atmosphere attack the ozone layer.

CHEMICAL REACTION: A process in which **elements** or **compounds** (the **reactants**) change to form different elements or compounds (the **products**). The change may be permanent or reversible. During a chemical reaction, **electrons** are transferred or shared between the reactants.

CHROMATOGRAPHY: A technique used to separate a **mixture**. The various types of chromatography all use a substance (known as the stationary phase) that takes up different parts of the mixture at different rates.

COLLOID: A type of **mixture**, similar to a **solution**, in which particles of one substance are distributed evenly throughout another. Colloidal particles are larger than those in a solution, but smaller than those in a **suspension**.

COMPLEX ION: A type of **ion** in which a central metallic **cation** is combined with surrounding **anions** or **molecules**. Iron and other **transition metals** form complex ions with water molecules.

COMPOUND: A pure substance in which **elements** are chemically combined in a definite ratio. In the compound water, H_2O, **atoms** of hydrogen and oxygen are bound together in the ratio 2:1. Compounds with **covalent bonding** generally consist of **molecules**. When solid, compounds with **ionic bonding** consist of **macromolecules**.

CONCENTRATION: The amount of a dissolved substance (**solute**) present in unit volume of **solution**. Molar concentration has units of **moles** per litre (mol l⁻¹ or mol dm⁻³). The units of mass concentration are kilograms per litre (kg l⁻¹ or kg dm⁻³).

COVALENT BONDING: A type of chemical bonding in which **electrons** are shared between the **atoms** involved. **Compounds** that exhibit this type of bonding are called covalent compounds.

CRYSTAL: A regular arrangement of **atoms**, **ions**, or **molecules** in a solid. This regular internal structure leads to a geometrically regular

REHYDRATING ANHYDROUS CRYSTALS

external shape. Sodium chloride crystals, for example, are cubic.

DECOMPOSITION: Any **chemical reaction** in which a **compound** breaks down into simpler compounds or **elements**. Many compounds decompose upon heating or **electrolysis**.

DEHYDRATING AGENT: Removes water from another substance in a **chemical reaction** called dehydration. Some dehydrating agents can remove hydrogen and oxygen in the ratio 2:1 to make water where there was none before.

DIATOMIC: Describing a **molecule** that is made up of two **atoms**. Hydrogen is a diatomic gas.

DISPLACEMENT REACTION: A **chemical reaction** in which one **atom**, **ion**, or **molecule** replaces another. Zinc displaces copper from a **solution** of copper(II) ions, Cu^{2+}.

DISTILLATION: Boiling a liquid to vaporize it, and then condensing the vapour back into a liquid in a separate vessel. Distillation is used to separate the **solute** from the **solvent** in a **solution**. A **mixture** of liquids with different boiling points is separated by fractional distillation.

DOUBLE DECOMPOSITION: A **chemical reaction** between two **salts** in which **ions** or **radicals** are exchanged, usually in **solution**.

EFFLORESCENT: Describing a substance that loses some or all of its **water of crystallization** to the air, forming a new, often powdery substance. If all the water is lost, the **anhydrous** form results.

ELECTRIC CHARGE: The property of a particle that gives rise to the electrostatic forces responsible for chemical activity, and for holding **atoms** together. Charge can be negative or positive. Like charges repel, and unlike charges attract.

ELECTROCHEMICAL CELL: A system consisting of an **electrolyte**, two **electrodes** (a **cathode** and an **anode**), and an external electric circuit. There are two basic types of electrochemical cell: the electrolytic cell, used in **electrolysis** and electroplating, and the voltaic cell, as found in household batteries.

ELECTRODE: A plate made from an electrical conductor, sometimes graphite but usually metal, for use in **electrochemical cells**. In a cell, one electrode is the **anode**, and the other is the **cathode**.

ELECTROLYSIS: A process in which a **chemical reaction** occurs as a result of electric current being passed through an **electrolyte**. **Decomposition** of **compounds** can be achieved by electrolysis.

ELECTROLYTE: A paste, liquid, or **solution** containing **ions** that conducts an electric current. The current is carried by **electrically charged** ions, which move towards the oppositely charged **electrode**. Sodium chloride solution and molten sodium chloride are both electrolytes.

ELECTRON: A particle carrying a negative **electric charge** that is found in all **atoms**. In a neutral atom, there are equal numbers of **protons** and electrons.

ELECTRON SHELL: A set of **orbitals** in an **atom**, where **electrons** may be found. The first shell, closest to the **nucleus**, holds up to two electrons in an s-orbital. The second shell has one s- and three p-orbitals, holding up to eight electrons, while the third shell, which also has five d-orbitals, can hold up to 18. Usually, shells are filled progressively from the first shell outwards. Across a period, from group 1 through group 18, empty

PREPARATION OF NITROGEN DIOXIDE GAS

orbitals up to the current shell are filled. Moving from a group 18 **element** to the next (group 1) element, a new shell is begun.

ELECTROPLATING: A process in which metal **cations** from an **electrolyte** are deposited as a thin layer onto the surface of a metal object that has been made the **cathode**. Many items, from spoons to car bodies, are electroplated.

ELEMENT: A substance containing **atoms** with the same **atomic number**. Every element has characteristic chemical properties. There are 92 naturally occurring elements on Earth.

ENDOTHERMIC: Describing a **chemical reaction** during which heat **energy** is taken in from the surroundings and converted to chemical energy. Endothermic reactions are usually accompanied by a drop in **temperature** of the substances and apparatus.

ENERGY: The ability to make something happen. Energy must be expended in order to do work. Although the total amount of energy in the Universe is constant, it can take many interchangeable forms. The two basic forms of energy are potential energy and **kinetic energy**. Potential energy is that energy stored in a system, while kinetic energy is the energy it possesses due to its motion. For example, a nitrogen gas **molecule** in air has potential energy stored in the bond between its **atoms**, and kinetic energy due to its motion through the atmosphere. The potential energy stored in systems made up of atoms, **ions**, and molecules is called chemical energy. During **electrolysis**, electrical energy can be used to overcome this chemical energy to **decompose** molecules.

ENZYME: A **catalyst** found in, or derived from, a living organism that increases the **rate** of a **chemical reaction**. Enzymes are highly specific, usually catalysing a particular step in a long and complex chain of reactions. Nearly all enzymes are **proteins**.

EQUILIBRIUM: A stable state in a **reversible reaction**. Such a reaction can be thought of as two simultaneous reactions (the forward and reverse reactions). The reactions are in equilibrium when they proceed at the same **rate**, so that there is no overall change. The equilibrium position determines the proportion of **reactants** to **products** in the reaction vessel.

EXOTHERMIC: Describing a **chemical reaction** during which chemical **energy** of the **reactants** is converted to heat energy and given off to the surroundings. Exothermic reactions are generally accompanied by a rise in **temperature**.

FILTRATION: A method for separating **suspensions**. The suspension is passed through a filter, often made of paper, which is perforated with tiny holes. The holes allow only small **molecules** through, while the larger suspended particles are retained by the filter.

HYBRIDIZATION: The averaging out of bonding **orbital** energies in an **atom**. For example, in carbon atoms as found in diamond, each carbon is joined to four others with bonds of equal **energy**. Each carbon atom normally has one s- and three p-orbitals used for bonding. In diamond, these two types of orbital are converted into four sp³ hybrid orbitals, each with the same energy.

HYDROCARBON: A **compound** containing the **elements** carbon and hydrogen only. Hydrocarbons are classed as **organic** compounds.

EXOTHERMIC BURNING REACTION

different colours form according to the **pH**. Indicators such as litmus solution and universal indicator are used in chemical analysis.

ION: A particle with **electric charge**, formed when an **atom** gains or loses **electrons**. A positive ion is called a **cation**, and a negative ion is an **anion**.

IONIC BONDING: A type of bonding in which **cations** and **anions** are held together by forces due to their **electric charges**. The **ions** form a **crystal** structure called a **macromolecule**. **Compounds** that contain such bonds are called ionic compounds.

ISOTOPE: One of the possible forms of an **element** that differ in their **nuclear** structure. Although all **atoms** of a particular element have the same number of **protons** in the nucleus, different numbers of **neutrons** may be present. Different isotopes of an element have the same chemical properties, but different **RAMs**. The element fluorine has four isotopes: fluorine-17 (with nine protons and eight neutrons), fluorine-18, fluorine-19, and fluorine-20 (with 11 neutrons).

KINETIC ENERGY: The **energy** that a particle or an object possesses due to its motion or vibration. The more mass an object has and the faster it moves, the more kinetic energy it possesses. Heat energy is the kinetic energy of the random motion of the **atoms**, **ions**, and **molecules** that make up matter.

MACROMOLECULE: Any molecule with an **RMM** greater than about 10,000. The term is often used to refer to **ionic crystals**, such as those of sodium chloride.

MIXTURE: Two or more pure substances (**elements** or **compounds**) that are mixed but not chemically combined. The components of a mixture can be separated by methods such as **chromatography** and **filtration**.

HYDROGEN BONDING: Weak bonding between some **molecules** that contain hydrogen **atoms**, caused by the uneven distribution of **electric charge** within the molecules. Hydrogen bonding is found in water, and is responsible for its relatively high boiling point.

COPPER CHLORIDE CRYSTALS

HYDROXONIUM ION: Also called a hydronium or oxonium ion. This is an **ion** with formula H_3O^+, which consists of a **proton** or hydrogen ion, H^+, associated with a water **molecule**, H_2O. Hydroxonium ions form in equal numbers with hydroxide ions, OH^-, when water splits into ions. In a **solution** of an **acid**, the concentration of H_3O^+ is higher than that of OH^-, while the opposite is true of an **alkaline** solution. The **concentration** of hydroxonium ions in solution is given according to the **pH scale**.

HYGROSCOPIC: Describing a substance that absorbs water from the air.

INDICATOR: A substance, usually based on natural plant material, whose colour changes according to the **acidity** or **alkalinity** of its environment. When added to the test **solution**, **complex ions** of

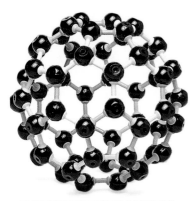

BUCKMINSTERFULLERENE

Solutions and **colloids** are two types of mixture.

MOLE: A unit of the amount of a substance, defined in terms of the number of particles present. One mole of a substance contains 6.02 x 10^{23} particles, and has a mass in grams equal to its **RAM** or **RMM** – so the mass of one mole of copper is 64.4 grams. The quantity 6.02 x 10^{23} mol^{-1} is Avogadro's constant.

MOLECULAR ORBITAL: A region within a **molecule** in which the **electrons** involved in **covalent bonding** are likely to be found. Molecular orbitals are formed by the overlap of the outer **orbitals** of the **atoms** that are bound together.

MOLECULE: The smallest unit of many **compounds**. It consists of two or more **atoms** held together by **covalent bonding**.

NEUTRON: A particle with no **electric charge** that is found in the **nucleus** of every **atom**, except those of the common **isotope** of hydrogen.

NOBLE GASES: The **elements** of group 18 of the periodic table. These elements are all gases at room temperature, and are very unreactive because they have filled outer **electron shells**.

NUCLEUS: The central, positively charged part of an **atom**, made up of **protons** and **neutrons**. The common **isotope** of hydrogen is the only type of atom that does not have neutrons in its nucleus.

ORBITAL: The region of space around an **atom**, an **ion**, or a **molecule** where **electrons** are likely to be found. In modern theory, electrons are seen to have only a probability of being in a particular region, and not to follow definite orbits. In an atom, the simpler types of orbitals are called s-, p-, and d-orbitals. Atomic orbitals hold up to two electrons each.

ORE: A mineral containing metal **atoms**, normally combined with atoms of oxygen or other **elements**.

ORGANIC: Describing a **compound** based on chains or rings formed by carbon **atoms**. These compounds are the basis of life as we know it. Organic chemistry is the study of such compounds.

OXIDATION: The removal of **electrons** from, or the addition of oxygen to, an **atom**, an **ion**, or a **molecule**. An **element** that is oxidized increases its **oxidation number**.

OXIDATION NUMBER: A positive or negative number that indicates whether an **element** has lost or gained **electrons** during a **chemical reaction**. When copper **atoms** lose two electrons to form doubly charged copper(II) ions, Cu^{2+}, the oxidation number of copper (initially 0) becomes +2, also given by the Roman numeral (II).

pH SCALE: A scale that indicates whether a **solution** is **acidic** or **alkaline**. The scale runs from 1 (strong acid), through 7 (neutral), to 14 (strong alkali). The pH value relates directly to the **concentration** of hydrogen **ions** in the **solution**.

PHOTOSYNTHESIS: A **chemical reaction** that occurs in green plants, during which the green pigment chlorophyll uses light **energy** to make **carbohydrates**.

PAPER CHROMATOGRAPHY

POLYMER: A large **molecule** that is formed by the joining of smaller molecules – units called monomers – in a reaction called polymerization.

PRECIPITATE: A solid substance formed by a **chemical reaction** taking place in a **solution**. Precipitates often form during **double decomposition** reactions.

PRODUCT: An **element** or **compound** that is formed in a **chemical reaction**.

PROTEIN: An **organic polymer** that contains carbon, hydrogen, oxygen, and nitrogen. Most proteins also contain sulphur.

PROTON: A particle with a positive **electric charge**, which is found in the **nucleus** of every **atom**. The charge on a proton is exactly the opposite of that on an **electron**.

RADICAL: An **ion**, normally consisting of two or more non-metals, that generally remains unchanged during a **chemical reaction**. An example is the carbonate ion, CO_3^{2-}.

RAM: Abbreviation for relative atomic mass. It is the mass of an **atom** of an **element** relative to $^1/_{12}$ of the atomic mass of the carbon **isotope**, carbon-12. RAMs are average values, weighted for the relative natural abundances of different isotopes of an element.

RATE OF REACTION: How quickly a **chemical reaction** proceeds. It depends upon various factors, including **temperature**, and may be increased by using a **catalyst**.

REACTANT: An **element** or **compound** that is the starting material of a **chemical reaction**.

REACTIVITY: A measure of the ease with which an **atom**, an **ion**, or a **molecule** reacts. **Elements** in groups 1 and 17 of the periodic table are generally the most reactive.

REDOX REACTION: Any **chemical reaction** involving the transfer of **electrons** (**reduction** and **oxidation**). Nearly all reactions can be seen as redox reactions.

REDUCTION: The addition of **electrons** to, or the removal of oxygen from, an **atom**, an **ion**, or a **molecule**. The **oxidation number** of an **element** that is reduced decreases.

REVERSIBLE REACTION: A **chemical reaction** in which the **products** can react to form the **reactants** once again. Many reactions that seem to proceed in only one direction are reversible reactions with an **equilibrium** position very close to the products.

RMM: Abbreviation for relative molecular mass. It is the sum of the **RAMs** of the **elements** that make up a **compound**. The RMM of water, H_2O, is 18, because the RAMs of hydrogen and oxygen are 1 and 16 respectively.

SALT: An **ionic compound** that is formed whenever an **acid** and a **base** react together.

SEMI-METAL: Also called a metalloid, an **element** that shows characteristics between those of metals and non-metals. Semi-metals are fairly good conductors of heat and electricity.

SOLUTE: Dissolves in a **solvent** to form a **solution**.

SOLUTION: An even **mixture** of two or more substances in which the particles involved are **atoms**, ions, or molecules. The **solvent** – a solid, liquid, or gas – dissolves one or more other substances (the **solutes**) to form a solution.

SOLVENT: Dissolves a **solute** to form a **solution**.

SPECTROMETER: Every **element** or **compound** produces a unique spectrum, or identifying pattern corresponding to **energy** levels in its **atoms**, **ions**, or **molecules**. A spectrometer is an instrument used to produce a spectrum for chemical analysis. Spectroscopes are spectrometers that use light. A mass spectrometer operates on a different principle – it is used to identify compounds by establishing their **RMMs**.

STP: Abbreviation for standard temperature and pressure. STP equals 0°C (32°F) and atmospheric pressure (101,325 Nm^{-2}).

SUSPENSION: A type of **mixture** in which particles, larger than those in a **colloid**, are unevenly distributed in a liquid or a gas. Suspensions can be separated by **filtration**. Muddy water contains soil particles in suspension.

TEMPERATURE: A measure of how hot or cold a substance is. The temperature of a substance is directly related to the average **kinetic energy** of its **atoms**, **ions**, or **molecules**.

TITRATION: A procedure in which a measured amount of one solution of known **concentration** is added to another solution, usually in order to determine the latter's concentration.

TRANSITION METAL: Those **elements** that are found in the d- and f-blocks of the periodic table. Most metals, including iron and copper, are transition metals.

WATER OF CRYSTALLIZATION: Water that is held in **crystals** of a **compound**.

PRECIPITATION REACTION

Index

Acknowledgments

Dorling Kindersley would like to thank:
The Hall School, South Hampstead, and Imperial College, London, for the use of their laboratories; University College, London.

Special thanks to Richard Orchard at The Hall School; Chris Sausman at Imperial College; Peter Leighton at University College; Patrick Rolleston at Kensington Park School.

The Author wishes to thank the editorial and design team at Dorling Kindersley for their dedication and scrupulous attention to detail in the production of this book. Their technical support was invaluable. Also, thanks to the person who convinced me that chemistry was the "central science".

Additional Editorial Assistance:
Jo Evans.

Picture research:
Sharon Southren.

Picture credits:
(t=top, b=bottom, c=centre, l=left, r=right)
The British Museum 43br; The British Petroleum Company Plc 33tl; British Steel Plc 41tl; Leonard Lessin/Peter Arnold Inc 53cl; NASA 49br; Philippe Plailly 16tr; Pictor 39 tl; Des Reid p29 cl; Science Photo Library/CNRI 39tr.